SCIENCE ADVENTURERS

FOSSIL HUNTERS

BY TRISTAN POEHLMANN

CONTENT CONSULTANT

Kevin Padian
Professor and Curator
Department of Integrative Biology and Museum of Paleontology
University of California, Berkeley

Essential Library
An Imprint of Abdo Publishing
abdobooks.com

ABDOBOOKS.COM

Printed in the United States of America, North Mankato, Minnesota.
092019
012020

THIS BOOK CONTAINS RECYCLED MATERIALS

Cover Photos: iStockphoto (front); Shutterstock Images (back)
Interior Photos: Robert Clark/National Geographic, 4–5, 8, 78, 82; Siphiwe Sibeko/Reuters/Newscom, 11; Red Line Editorial, 13; The Natural History Museum, London/Science Source, 14–15, 23, 70–71; Martin Kemp/Shutterstock Images, 16–17; Dominic Gentilcore/Shutterstock Images, 20; Marcio Jose Bastos Silva/Shutterstock Images, 21; Michael Rosskothen/Shutterstock Images, 24; Barbara Barbour/Shutterstock Images, 26–27; Photo12/UIG/Universal Images Group/Getty Images, 30; Top Photo Group/Newscom, 32–33; Terrence Jennings/Polaris/Newscom, 35; Shutterstock Images, 36, 66–67; mauritius images GmbH/Alamy, 38–39; Chicago Photographer/Shutterstock Images, 43; Natursports/Shutterstock Images, 46–47; Pablo Blazquez Dominguez/Getty Images News/Getty Images, 50–51; James King-Holmes/Science Source, 55; Achmad Ibrahim/AP Images, 56–57, 59; AP Images, 63; The Sydney Morning Herald/Fairfax Media/Getty Images, 65; Roger Harris/Science Source, 75; Sputnik/Alamy, 76–77; Steve Jennings/Breathrough Prize/Getty Images Entertainment/Getty Images, 84–85; Javier Trueba/MSF/Science Source, 87; Paul Nicklen/National Geographic, 88–89, 92–93; Alfredo Estrella/AFP/Getty Images, 96–97; Paleontologist Natural/Shutterstock Images, 98

Editor: Arnold Ringstad
Series Designer: Laura Graphenteen

LIBRARY OF CONGRESS CONTROL NUMBER: 2019941973

PUBLISHER'S CATALOGING-IN-PUBLICATION DATA

Names: Poehlmann, Tristan, author.
Title: Fossil hunters / by Tristan Poehlmann
Description: Minneapolis, Minnesota : Abdo Publishing, 2020 | Series: Science adventurers | Includes online resources and index.
Identifiers: ISBN 9781532190346 (lib. bdg.) | ISBN 9781532176197 (ebook)
Subjects: LCSH: Fossils--Juvenile literature. | Archaeologists--Juvenile literature. | Scientists--Juvenile literature. | Discovery and exploration--Juvenile literature. | Adventure and adventurers--Juvenile literature.
Classification: DDC 560.9--dc23

CONTENTS

CHAPTER ONE
INTO THE DEEP TOMB 4

CHAPTER TWO
ALONG THE JURASSIC COAST 14

CHAPTER THREE
UP THE FLAMING CLIFFS 26

CHAPTER FOUR
DOWN THE LONG SINKHOLE 38

CHAPTER FIVE
INTO THE PIT OF BONES 46

CHAPTER SIX
UNDER THE SERPENT ISLAND 56

CHAPTER SEVEN
OVER THE OCEAN OF ICE 66

CHAPTER EIGHT
UP THE SIBERIAN MOUNTAIN 76

CHAPTER NINE
INTO THE FLOODED CAVE 88

ESSENTIAL FACTS 100
GLOSSARY 102
ADDITIONAL RESOURCES 104
SOURCE NOTES 106
INDEX 110
ABOUT THE AUTHOR 112

INTO THE DEEP TOMB

The heat of the South African sun faded as Alia Gurtov descended into the mouth of the cave. Down the ladder she went, into the dim cavern beneath the surface. The beam of Gurtov's headlamp swept across the ancient stone. The cave floor sloped under her feet. Ahead she could see where the chamber narrowed into a rocky tunnel. Following the guide rope that marked the way, she crawled forward into the dark.

The excavation team called the tunnel the Post Box—a low passageway that banged up their elbows and deposited them into the next chamber like letters through a mail slot. After that, the going got tough. Halfway through her journey underground, Gurtov approached Superman's Crawl. She entered the narrow tunnel arms first, as if she were flying. The opening was less than ten inches (25.4 cm) high.[1] Gurtov's hands scrabbled and her

Expeditions to find fossils sometimes involve climbing deep below Earth's surface.

shoulders scraped against the rock as she wriggled forward inch by inch. After 15 feet (4.6 m), Gurtov pulled herself out the other side.[2]

Here, the cave widened and seemed to soar upward. Craggy outcroppings rose high above Gurtov in a series of sharp peaks. This was the Dragon's Back, and she had to climb it. Belted into a safety harness, Gurtov began to scale the beast. Her headlamp guided her through the dark as she gripped rough handholds and clambered over jagged edges. Reaching the top, she met her next challenge, the Chute—a bruising squeeze all the way back down.

FIRST FOSSIL HUNTERS OF AFRICA

People across Africa collected fossils long before outsiders began to visit on fossil-hunting expeditions. More than 6,000 years ago, Egyptians used fossils for jewelry and religious rituals. Moroccan people recorded the discovery of fossilized elephant bones more than 2,000 years ago.[4] In Lesotho, ancient cave painters drew pictures of the fossilized dinosaur footprints found in their area.

Nearly at her journey's end, Gurtov pressed on. The Chute was a straight shot down into the final chamber, but it could barely be called a tunnel. It wasn't much more than a tight crevice split through the rock. Gurtov wedged herself into it, turning her limbs and head to become as flat as possible. Steadily, she lowered herself down almost 40 feet (12 m) in the pitch-black dark. Sharp rocks scratched her knees and elbows. Stone spikes scraped her face. At last she felt her feet hit the top rung of a ladder. Emerging from the crevice, Gurtov turned to survey Dinaledi Chamber, a cavern she had journeyed halfway around the world and 300 feet (90 m) underground to find.[3]

THE EXPEDITION

Dinaledi Chamber lies in a remote part of the Rising Star cave system near Johannesburg, South Africa. In 2013, amateur cavers explored the area. Entering the far reaches of the tunnels, they caught a glimpse of a human skeleton. They reported the sighting to Lee Berger, a paleoanthropologist at a nearby university. Immediately, Berger organized an expedition to recover the ancient bones.

Dinaledi Chamber means "Chamber of Stars" in the Sesotho language spoken in South Africa.

The Rising Star cave is in a region known as the Cradle of Humankind. For years, fossilized bones of ancient hominin species have been found throughout the area. But the difficulty of reaching these particular bones posed a new challenge. Berger recruited a team of experienced, small-statured paleontologists who had worked in caves before. The excavation team, all women, joined the expedition at a tent camp near the entrance to Rising Star. The cave was mapped in 3D using drones, laser scanners, and a program that recorded fossil positions. On a live video feed, Berger's team watched from the aboveground command center—a large green tent—as Gurtov and her fellow scientist-cavers navigated the treacherous tunnels.

Nearly 100 feet (30 m) straight down through thick layers of limestone, the team began its careful work of excavation.[5] In a churchlike chamber lit only by headlamps, Gurtov's team sweated in the still, damp air, their legs cramping as they unpicked a puzzle of bones.

Narrow passageways become laboratories and command centers for cave-exploring scientists.

For up to eight hours at a time, the scientists crouched barefoot so that they could feel if they accidentally stepped on the precious bones. What they recovered from Dinaledi Chamber was the largest collection of hominin fossils ever found on the continent of Africa.

ANCIENT BONES

Near the end of 2013, the Rising Star Expedition went public with its finds. In less than a month's excavation, the team had dug a one-square-yard (0.84 sq m) trench using toothpicks and delicate brushes. The area was just a fraction of the cave floor. But already they had recovered more than 1,500 fossilized bones and teeth belonging to ancient relatives of modern humans.[6] It was a stunning amount of material. All the fossils needed to be carefully cleaned, studied, and analyzed in a laboratory.

UNDERGROUND ASTRONAUTS

Lee Berger, leader of the Rising Star Expedition, called the team members who journeyed deep into the cave system "underground astronauts." Their work was physically and mentally difficult. They risked their lives to recover priceless scientific information. "I watched these remarkable scientists from the command centre on infrared cameras," he said. "The remoteness of where they were working . . . [and] the ever present danger really did [make] it feel like a NASA mission."[8]

Excavation would continue, but the team of paleoanthropologists turned their focus to the lab. Boxes of fossilized bones, recorded and tagged with identifying information, were delivered to a vault at Berger's university. Over the next two years, scientists from all around the world collaborated in their analysis. Reconstructing the hominin skeletons, they found that the bones came from at least 15 different individuals.[7] They were adults, children, elderly people, and newborn babies. By studying their fossilized bones, scientists were able to discover much about how these ancient people looked, moved, and behaved.

The people of Dinaledi Chamber resembled modern humans in a few ways. They stood about five feet (1.5 m) tall and walked upright. Their faces were humanlike. But their body proportions revealed intriguing differences. Their broad chests, wide hips, and slender legs indicated that they could walk long distances. Their strong hands and long, curved fingers and toes showed they were skilled at climbing trees. Though their brains were small—about the size of an orange—they appeared able to make and use tools.

Paleoanthropologists compared the Dinaledi Chamber fossils to hominin fossils from other excavation sites. The results puzzled them. The Dinaledi bones showed a mix of different characteristics. They did not perfectly match any of the known hominin species. But there was a good reason for that. In 2015, the Rising Star team announced that the Dinaledi people represented a totally new species of hominin.

ORIGIN OF SPECIES

In 1961 in a cave in Morocco, a group of miners found a fossilized human skull. The skull wasn't considered very important until 2004. That's when a curious French scientist began an excavation at the cave, called Jebel Irhoud. He found more fossilized bones. When he had the bones dated, he discovered that the cave held the oldest *Homo sapiens* ever found. The Jebel Irhoud skull is 300,000 years old.[9] It is the earliest known fossil of the human species.

A NEW SPECIES

People around the world were excited by the discovery of a previously unknown hominin species. The species was named *Homo naledi*, after the Dinaledi Chamber where the fossils were found. But for years the Rising

Star team struggled to calculate how old its fossils were. How long ago had *Homo naledi* lived?

By 2017, several labs had independently tested fossilized teeth and rock from the cave. The labs used different methods to measure the concentration of radioactive elements in the fossils. These elements decay at a steady rate, which can be measured and used to calculate a fossil's age. The labs' results overlapped to show that the bones were 236,000 to 335,000 years old.[10] This date range revealed that *Homo naledi* lived during the same time period as early *Homo sapiens*. Although the people from Dinaledi Chamber were a different species than early humans, they walked the earth alongside modern humans' ancestors.

Homo naledi also practiced a surprisingly human tradition. Dinaledi

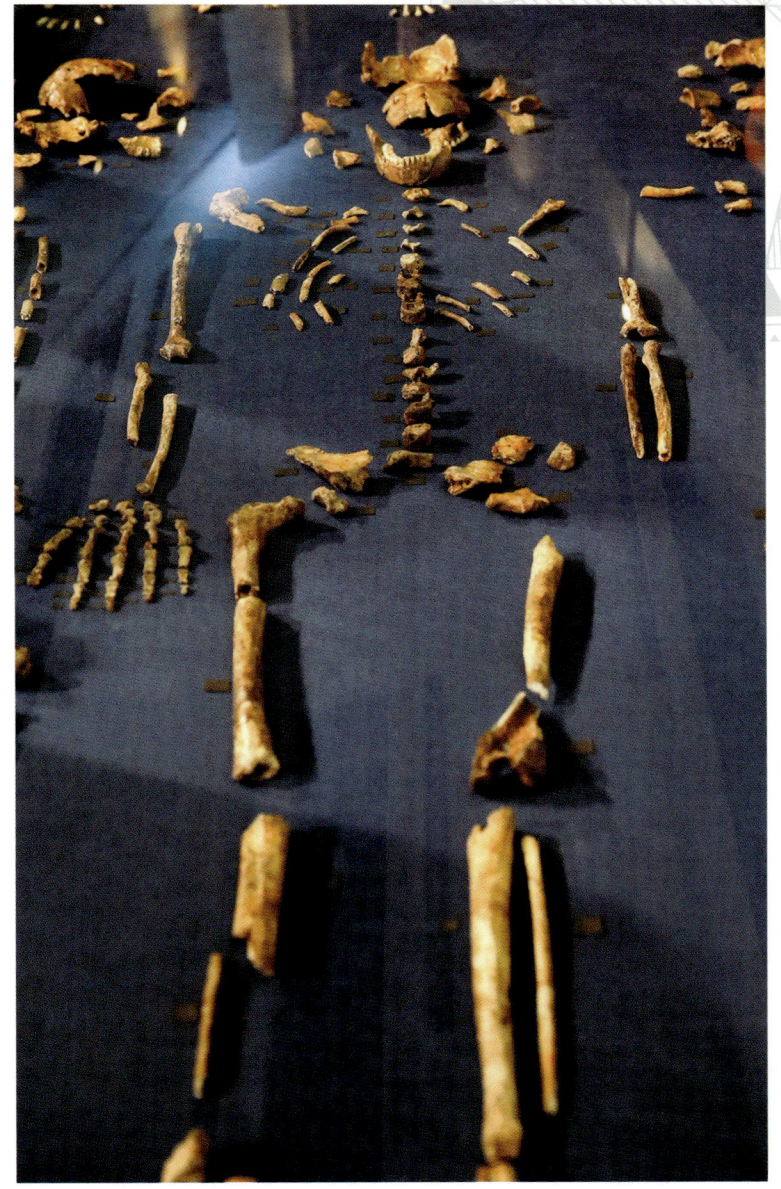

Scientists reconstructed *Homo naledi*'s skeleton from the pieces they uncovered.

Chamber had once been a tomblike place where the species placed its dead. The practice proved to the Rising Star team that the species had intelligence. But how these prehistoric people reached Dinaledi Chamber is still a mystery.

ONGOING WORK

While *Homo sapiens* evolved and survived, *Homo naledi* died out. Human evolution includes events that often look, in retrospect, like failed experiments. But as John Hawks, a paleoanthropologist on the expedition, explained, "We have all these things we think of as human. . . . Walking upright is human, a large brain is human. . . . But all of these things happened at different times in different ancestors. The package we think of as human did not appear simultaneously."[11] Evolution is not a straight line from point A to point B.

The Rising Star team will continue to study *Homo naledi* both in the cave and in the lab. Wherever it happens in the world, fossil hunting is difficult but rewarding scientific work. An expedition starts with long days chiseling rock and brushing soil. Excavators record fossil positions with GPS, document their size with measuring tape, and preserve fragile fossils with special glue. In the lab, they carefully clean and reconstruct broken fossils for study. Hunting for fossils is just the beginning of this exciting journey of discovery.

Homo sapiens means "wise human" in Latin. *Homo naledi* means "star human," from Latin and Sesotho.

TIMELINE OF HOMININ SPECIES

HOMO HABILIS : 2.4 million–1.4 million years ago

HOMO RUDOLFENSIS : 1.9 million–1.8 million years ago

HOMO ERECTUS : 1.89 million–143,000 years ago

HOMO HEIDELBERGENSIS : 700,000–200,000 years ago

HOMO NEANDERTHALENSIS (NEANDERTHALS): 400,000–40,000 years ago

HOMO NALEDI : 335,000–236,000 years ago

HOMO SAPIENS (MODERN HUMANS): 300,000 years ago–present

HOMO FLORESIENSIS: 100,000–50,000 years ago[12]

| 2.5 | 2 | 1.5 | 1 | .5 | 0 |

Millions of Years Ago

ALONG THE JURASSIC COAST

It was 1811. The sharp wind yanked at 12-year-old Mary Anning's bonnet and dress as she paced along the sheer sea cliffs of England's southwest coast. High above her, the family home stood small and ramshackle against the gray sky. Peering closely at the walls of rock, she made her way toward the spot her older brother had shown her. Last year, he had discovered a skull there, and he had told her how it fascinated all the tourists and collectors who visited the market stall where their mother sold fossils. Anning didn't know what kind of strange beast the long-snouted skull came from, but she was determined to find the rest of the animal. A complete skeleton would be worth a lot of money to the wealthy fossil collectors who purchased their finds.

Mary Anning spent most of her life as an avid fossil hunter. ▷

For months, Anning had been waiting for winter to turn to spring and the harsh sea storms to quiet down. Now that the cliffs were dry and the rockfalls and mudslides had ended, she could begin to excavate. Reaching the spot, she dropped her basket of tools and squinted at the rock face. She could just make out a faint outline of bone in the crumbling limestone. The winter storms had worn away at her brother's chisel marks, revealing the skeleton beneath. Anning set to work with her own chisel and hammer.

For months, Anning hammered and scraped all day long on the windy cliffs. Townspeople visited the spot to watch the beast emerge beneath her chisel. It was clear to them that the bones were those of a monster. The huge, long body seemed to continue on forever. Doggedly, Anning kept chipping away at the limestone that encased the animal. After months

The Jurassic Coast stretches for dozens of miles along southern England.

of labor, she finally uncovered its tail end. The fossilized skeleton stretched an astonishing 17 feet (5.2 m) across the cliff.[1]

FROM SOUVENIRS TO SCIENCE

Fossil hunting wasn't always about scientific discovery. Before scientists understood evolution, people believed many different things about plants and animals preserved in rocks. Some thought they were totally separate from living organisms. Others thought these living things had migrated from other places in earlier times. And some thought they were the remains of ancient creatures or dragons. Collecting fossils was a popular hobby, particularly for wealthy people who could afford to take vacations by the sea. The odd fossils unearthed on the southwest English coast quickly became desirable souvenirs to display in family homes. Mary Anning's hometown of Lyme Regis eventually became a world-renowned site on the so-called Jurassic Coast, where fossil hunters collected marine fossils from the Jurassic period.

Anning was a child when she discovered her first important fossil. She had grown up fossil hunting with her father. The family sold small fossils to tourists from their struggling cabinetry shop. But one day Anning's

THE OLDEST ANIMAL ON EARTH

For many years the earliest forms of animals were mysterious. Then, in 2018, scientists proved the existence of a life form that was 558 million years old.[2] In a remote part of Russia, a worker hanging over the edge of a sandstone cliff recovered a pancake-shaped fossil. Incredibly, it contained preserved organic matter. Lab testing proved the organic matter was animal fat. The ancient fossil held proof of an extremely early animal, a jellyfish-like marine creature called *Dickinsonia*.

father fell from the high cliffs and was badly injured. He died in 1810, and the family sank further into poverty. Fossil hunting became their only source of income. Finding a huge, complete skeleton meant that a wealthy man or a scientist might pay the family enough money to live on.

Anning's mother sold the gigantic fossil to a businessman for 23 £ (approximately $1,300 in today's money), which helped the family, but it wasn't enough to pay off their debts. The businessman sold it to a collector, and scientists in London debated whether the fossil was an ancient crocodile or a deep-sea fish. Years later, scientists named it *Ichthyosaurus*, meaning "fish lizard." In fact, the fossil was a species of marine reptile that lived between 194 and 201 million years ago.[3]

THE FOSSIL DEPOT

As Anning grew older, her skills improved. She continued to hunt fossils along the cliffs and beaches. The work was dangerous, and she experienced close calls, including a disaster that killed her beloved dog. When her older brother began to work as an upholsterer, she continued the family business herself. Anning's Fossil Depot sold ammonites, belemnites, and other common local finds to tourists and collectors.

FOSSILIZED FECES

Mary Anning was the first fossil hunter to study and identify coprolites—the fossilized feces of ancient animals. At the time, no one knew what the oddly shaped fossils were. Anning decided to split one open and use her knowledge of anatomy to study its inside. But instead of a skeleton, what she found was a mass of fragmented fish bones, scales, and seashells. She inferred that coprolite fossils were in fact animal droppings. These fossils had the potential to reveal a whole new set of information about an ancient animal's diet, anatomy, and environment.

SCIENCE CONNECTION

BODY FOSSILS AND TRACE FOSSILS

Body fossils are preserved parts of an organism's body. The simplest form is total preservation of a body, such as an insect caught in tree sap or a woolly mammoth frozen in ice. In arid climates, bodies can be preserved when they dry out. Most of the fossils commonly seen, however, form during the creation of sedimentary rock. A body is covered and preserved under layers of sediment, sealing it inside the rock. Over time, groundwater brings in sedimentary particles that fill the microscopic spaces in the tissues.

Often minerals alter or replace the body as it decays, sometimes almost completely.

Trace fossils are formed from traces an organism leaves behind. They can be tracks such as footprints, trails of burrowing animals, or digested material such as feces. Tracks and other imprints are preserved in mud or wet sand by layers of sediment that solidify into rock. Feces and other evidence of feeding are preserved in sediment and replaced by minerals as they decay, leaving behind hard casts of the material.

A trace fossil of a dinosaur footprint

Although Anning learned to read and write as a child, she had no formal education. She worked hard to teach herself geology and anatomy by reading scientific papers. She studied the language and methods of science, excavation and cleaning of delicate fossils, and proper reconstruction of fragmented skeletons. She learned to identify ancient species, mount them for display, and document them with anatomical drawings. She continued to live in poverty, but her professional reputation grew. Scientists began to visit Anning's Fossil Depot to discuss her unique specimens.

STRANGE CREATURES

In 1823, when Anning was still a young woman, she made a stunning discovery that amazed the European scientific community. Excavating the cliffs at Lyme Regis, she recovered another complete skeleton of an ancient animal no one had ever seen before. The anatomy of the fossilized creature was so strange that many scientists doubted the accuracy of Anning's illustration, believing that such a fossil must be a fake. Her fossil was taken to London to be analyzed by the Geological Society, but Anning herself was not invited. After much debate, the scientists of the Geological Society concluded that the fossilized skeleton was genuine. Named *Plesiosaurus*, meaning "close to lizard," the animal made Anning famous.

"[Mary Anning is] in the habit of writing and talking with professors . . . and they all acknowledge that she understands more of the science than anyone else in this kingdom."[4]

—Lady Harriet Silvester, a tourist who visited Anning in 1824

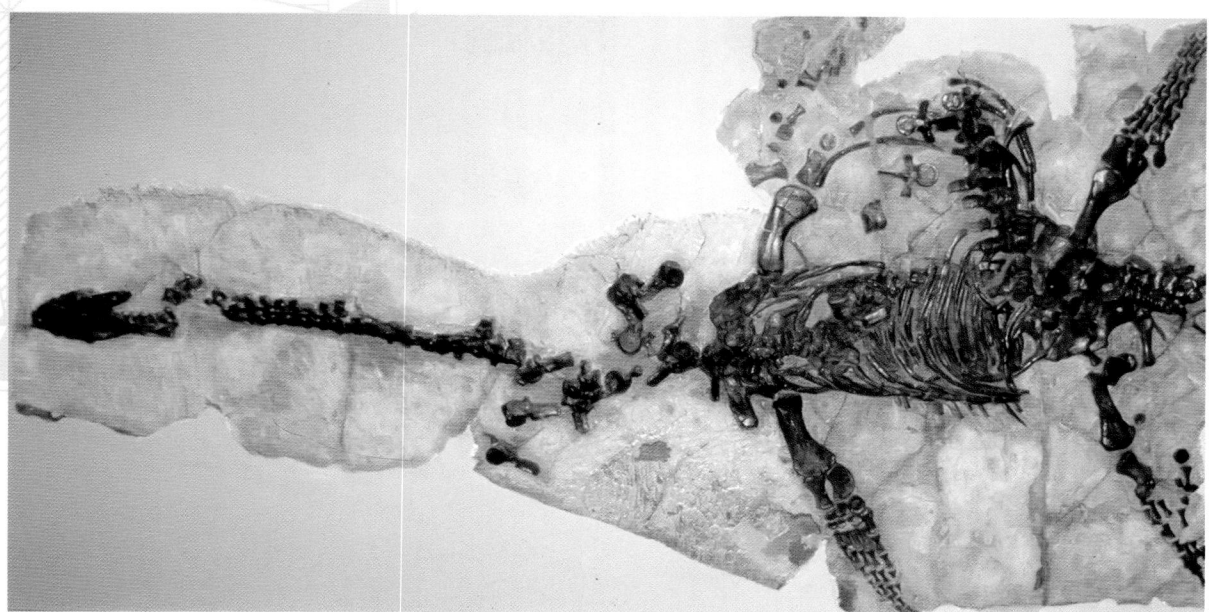

The *Plesiosaurus* fossil Anning uncovered is preserved at the Natural History Museum in London, England.

Though Anning became a respected fossil expert and was often consulted by scientists, they rarely publicly acknowledged her work. She was not allowed to join the Geological Society, and she was never considered a scientist herself. However, she drew many visitors to Lyme Regis, and her fascinating discoveries continued. In 1828, she excavated another astonishing skeleton with wings and a tail—the first pterosaur ever found in England. The famous paleontologist Richard Owen eventually named it *Dimorphodon* after its two kinds of teeth, large and small.

Anning's contributions to early paleontology were significant, and many of her finds became the basis of later scientific work. One scientist who purchased fossils from her

Dimorphodon had a large head for its body size.

later wrote, "This lady, devoting herself to Science, explored the frowning and precipitous cliffs . . . and rescued from the gaping ocean, sometimes at the peril of her life . . . specimens which originated . . . ingenious theories."[5]

Anning's skill at fossil hunting was legendary. She discovered countless ancient life forms hidden in the limestone cliffs of her hometown. Her meticulous excavations could take months of work. However, fossil hunting is not simply a matter of collecting bones. The careful process of scientific analysis is perhaps even more critical.

LOST AND FOUND

In 2015, a metal detectorist hunting for old coins and musket balls on Lyme Regis beach made an amazing find. It was a small metal disc about the size of a quarter. One side was stamped "Mary Anning, 1810" and the other "Lyme Regis, Age 11."[6] The discovery is now on display at the local museum, where experts verified its age. The museum director believes Anning's father made it for her as a birthday present shortly before he died. It's possible that she lost the token more than 200 years ago while fossil hunting on the cliffs above the beach.

CHAPTER THREE

UP THE FLAMING CLIFFS

As the sun began to set over the chilly desert, the high sandstone dunes lit up in shifting shades of red, orange, and gold, as if they were aflame. Roy Chapman Andrews ordered the caravan of motorcars and camels to slow down as he spotted a group of round tents ahead of them. Andrews's expedition had been on the same trail for some time, and settlements in the harsh Gobi Desert were rare. Over the past few months, the caravan had traveled thousands of miles and climbed 5,000 feet (1,500 m) above sea level.[1] Andrews needed to ask local people which trail would lead his expedition out of Mongolia for its return to China.

The Flaming Cliffs of Mongolia have long been a popular site for fossil hunters.

DUNE DWELLERS

The first fossil hunters to excavate the Flaming Cliffs lived in the area approximately 20,000 years ago. After ancient humans had migrated out of Africa and across Southeast Asia, a group of Stone Age people settled in the region of contemporary Mongolia. Known today as the Dune Dwellers, these people collected and used dinosaur fossils. During Roy Chapman Andrews' excavations in the 1920s, fragments of dinosaur egg fossils were found alongside burned animal bones and other evidence of human occupation. The Dune Dwellers lived millions of years after dinosaurs became extinct, so the location of these fragments suggests that they collected these fossils intentionally.

While the caravan waited for Andrews' instructions, the expedition's photographer wandered toward the edge of a nearby sand dune. Gazing across a wide valley of red sandstone, he decided to climb down and spend the short break searching for fossils. The slope was steep and difficult to navigate, but within minutes he stumbled upon an amazing sight. Perched on the peak of a rock, almost as if it were displayed in a museum, was the fossilized skull of a reptile. Carefully, the photographer picked it up, turned, and climbed back toward the line of dusty black motorcars above.

When Andrews returned to the caravan, he found the expedition's scientists poring over the fossil excitedly. None of them had ever seen anything like it. Andrews called a halt for the evening, and in the remaining light the men combed the bluff for signs of other fossils. By the time the sun set on the pitched tents and red cliffs, the expedition had collected an amazing array of dinosaur fossils.

GRAND EXPEDITIONS

With his foreign expeditions and wild adventure stories, Roy Chapman Andrews has been called the inspiration for the movie character Indiana Jones. Whether or not he really inspired the movie series, Andrews's 1920s expeditions to Mongolia made the Gobi Desert famous for its dinosaur fossils. The Flaming Cliffs, as Andrews named the site, became well known among paleontologists.

The griffin, an ancient Greek mythological beast that was half lion and half eagle, may have been based on stories from traders who had seen *Protoceratops* fossils in Mongolia.

Over time, however, Andrews became known for his mistakes as well as his discoveries. His expedition to Mongolia was originally intended to find proof of the hypothesis that ancient humans evolved in Central Asia. This hypothesis was incorrect. Instead, Andrews found rare evidence of dinosaur life in the Cretaceous period, an outcome he had not predicted. After stumbling upon the wealth of dinosaur fossils, Andrews returned the next year to excavate the Flaming Cliffs. Funded by the American Museum of Natural History, his numerous expeditions helped the museum amass an impressive fossil collection of previously unknown dinosaurs.

In the summer of 1923, a member of Andrews's expedition discovered a group of what seemed to be fossilized eggs on a small ledge, nearly buried in sand. At first, Andrews refused to believe that fragile items such as eggs would have been preserved, but he later agreed that the fossils were eggs. Many years before, French scientists had found spheres that they thought could be dinosaur eggs, but the evidence was not conclusive. Andrews' specimens were long, ovoid, and looked very much like some living reptile eggs. The fact that about a dozen were found together suggested that they were from a nest, even though there were no remains of plant material that could have helped form the nest. Even

Andrews's team uncovered a wide assortment of dinosaur fossils in his expeditions to Mongolia.

more amazingly, a fossilized skeleton of a somewhat birdlike dinosaur was found on top of the nest. Andrews interpreted it to be in the process of robbing the nest in order to eat the eggs. The species was named *Oviraptor*, meaning Egg Stealer. It is not clear why Andrews did not assume the dinosaur was guarding its own nest. Instead, he believed the nest belonged to a *Protoceratops*, a small horned dinosaur also found in the vicinity. His error was not discovered until 70 years later.[2]

RETURN TO THE DESERT

In the 1990s, the American Museum of Natural History funded a new series of paleontological expeditions to Mongolia, this time in collaboration with the Mongolian Academy of Sciences. In 1993, a scientist on the fourth expedition made an extraordinary discovery that contradicted established paleontological knowledge.

At a site near the Flaming Cliffs, Mark Norell found a nest of fossilized eggs that contained several fossils of embryonic dinosaurs. These eggs shared unusual characteristics in common with those found in 1923. They were about twice the size of a chicken egg and elongated, with a textured, crinkled shell. But the exterior was where their similarities ended. Inside the shells, Norell's dinosaur embryos looked nothing like Andrews's *Protoceratops*. The tiny, delicate skulls had long snouts and sharp little teeth rather than the expected toothless and protruding curved jaw of the *Protoceratops*.

HUNTING FOR FOSSILS

Despite huge advances in high-tech analysis of fossils in the last several decades, the basic methods of fossil hunting remain the same. They are surprisingly low tech. Paleontologists begin work at a site by prospecting. They scan the ground and their surroundings for the color of bone, which is often different from the surrounding terrain, or for signs of recent erosion. Areas that have been eroded by wind, water, or other natural wearing processes are the best places to find newly exposed fossils preserved inside rock or soil.

When a specimen is located, paleontologists first use brushes, awls, hand scrapers, and dental tools to determine the boundaries of the area that might contain fossils. When they believe they have identified the specimen's limits, they dig a trench around it. Without exposing or removing the fossil material, they form a cushion of wet paper on its surface. They then wrap the entire section of rock in strips of burlap soaked in plaster, much like a cast for a broken leg. When this material dries, paleontologists remove the section from its underlying pedestal of rock. Because a crucial piece of information about a fossil is exactly where it was found, paleontologists take detailed notes, photos, and GPS data on the site, marking the fossil's exact position on a topographic map.

The scientists were shocked. This find implied that for decades paleontologists had misattributed the species of this type of dinosaur egg. After the expedition, when the fossilized embryos were studied in a laboratory, they were found to be still tightly curled in the fetal position and almost perfectly intact. By carefully examining one of the fossilized skeletons with a needle-sized instrument, scientists determined that the shape and contour of the skull confirmed the species: *Oviraptor.*

Correcting a 70-year-old mistaken interpretation of dinosaur behavior had other unexpected scientific benefits. Paleontologists learned that *Oviraptors*, like birds, were brooding animals that lay on top of their nests, protecting or incubating their eggs. Norell's team also discovered that the jawbones of the embryonic skeletons had not yet fused together. They argued that this showed an evolutionary similarity to the embryonic development of modern

More preserved *Oviraptor* eggs have since been found at sites in China.

Dinosaur fossils are often stolen from remote areas like the Flaming Cliffs and sold for huge profits—an illegal practice that destroys scientific evidence. Fossils for sale that come from China, Mongolia, Argentina, or most other foreign countries have been exported illegally, regardless of what their sellers might say about having official paperwork.

THE FIGHTING DINOSAURS

Some of the most astonishing fossils ever found, the so-called fighting dinosaurs, were discovered only a few miles west of the Flaming Cliffs. In 1971, a paleontological expedition uncovered the remains of a battling *Velociraptor* and *Protoceratops*. The two dinosaurs had been buried alive in the middle of a violent fight approximately 80 million years ago.[3] It appears that the dinosaurs were buried by a collapsing sand dune or an unexpected, fierce sandstorm that immobilized the pair mid-combat. Over time, they became fossilized in position, the *Velociraptor* clawing the *Protoceratops*'s neck while the *Protoceratops* bit the *Velociraptor*'s front leg.

birds. Studying the fossilized skeleton of a brooding *Oviraptor* further revealed that the dinosaur's position on top of its nest suggested that its forelimbs were feathered. In the 1970s, John Ostrom of Yale University first showed that birds are descended from small carnivorous dinosaurs. Since then, discoveries by Norell's team and others have shown that many features associated with birds, such as nest-building and brooding, first evolved in their extinct dinosaurian relatives.

SAVING TREASURES

Bolortsetseg Minjin, a Mongolian paleontologist who took part in the expeditions of the 1990s, believes that the people living near the Flaming Cliffs should have access to the fossils discovered there. In the past, scientific education about dinosaurs had been very limited in Mongolia, even as foreign expeditions excavated and exported the region's

Bolortsetseg has played a key role in bringing Mongolian fossils back to their country of origin.

Museums in Mongolia showcase the nation's rich collection of fossils.

priceless paleontological artifacts. "Once a fossil left the country," Bolortsetseg explained, "knowledge left with it."[4]

Bolortsetseg founded the Institute for the Study of Mongolian Dinosaurs and began working with the Mongolian government to repatriate dinosaur fossils. In 2015, she worked with American colleagues to bring replicas of fossils in the American Museum of Natural History to Mongolia. She used a mobile museum to travel to rural areas and teach school children about their national treasures. Most students she talked to had never heard of dinosaurs or were not sure whether the animals were real or mythical. In the future, Bolortsetseg plans to build a museum at the Flaming Cliffs, making local dinosaur fossils a permanent part of Mongolian science education.

In the past, it was common for fossils to be excavated and exported from developing countries on a large scale. At the time, these countries usually had no museums or institutions able to preserve and study the specimens. Many had no interest in fossils. As the colonial eras ended and these nations grew, they focused more on preserving their fossil specimens. Scientists from other countries worked with them to carry out new expeditions, train people in science and research, and establish museums. Many fossils have been returned to their home countries. In other cases, fossil casts have been returned. Established museums also share original specimens and casts with new museums in developing countries.

DOWN THE LONG SINKHOLE

Deep inside a damp limestone cave, more than 30 feet (9 m) underground, a group of cave explorers glanced upward in surprise. High above their heads, through a collapsed sinkhole, weak sunlight beamed into the cavern. The dim light illuminated the path ahead of the group—a tunnel that appeared to continue on into endless dark. The group of explorers resumed their journey, making their way carefully into a tunnel that carried them down deeper into the earth.

Walking on the cave floor inside the tunnel was difficult. Knobby mineral growths and stalagmites covered ancient piles of animal bones, cementing them to the ground in tangled lumps. Once in a while, a side chamber

The rocky landscapes, sinkholes, and caverns of Italy hide evidence of the region's prehistoric inhabitants.

branched off of the 200-foot (60 m) main passageway. The explorers looked inside each one, headlamps sweeping across the candlelike dripstones covering the rock face.

Almost 60 feet (18 m) underground, the group reached the end of the tunnel.[1] The only remaining opening was one last small side chamber. Peering inside the dark, narrow shaft, the cavers faced a disturbing sight. Buried deep in a matrix of bulbous mineral growths, an upside-down human skull stared back at them.

ANCIENT MIGRATION TO EUROPE

Scientists estimate that an ancient hominin species began moving out of Africa around 2 million years ago. This hominin species, which has not yet been definitively identified, eventually migrated across southern Eurasia. The species made it as far as East Asia and Southeast Asia by 1.6 million years ago. Later waves of migration from Africa by other hominin species mixed with the earlier waves.

The groups of hominins that settled in Eurasia did not begin to migrate toward Western Europe until climate changes occurred about 900,000 years ago. Hominin species followed the mammals they hunted, first settling in southern regions such as contemporary Italy and Spain. Within 100,000 years, as climate changes continued, hominins migrated into northern regions of Europe, settling in areas such as contemporary Germany.[2]

TRAPPED UNDERGROUND

A few miles outside the town of Altamura in southern Italy, a small opening splits a rocky limestone hill. It marks an ancient sinkhole that provides access to Lamalunga Cave below. The region is known for its caves and sinkholes. These structures have slowly turned the ground into a Swiss cheese–like maze of underground tunnels and seasonal creeks.

The caves are the result of rain that has eroded channels into the limestone landscape, creating a watershed system with cracks that allow rainwater to drain into and continually erode the underground tunnels. Sinkholes occur when surface erosion meets erosion underground and the limestone suddenly collapses. This process has been going on since before ancient hominins roamed the landscape.

In 1993, the group of cave explorers discovered a fossilized hominin skeleton deep inside Lamalunga Cave. They reported their find to scientists at the nearby University of Bari. Researchers from the university hypothesized that the hominin had fallen through the sinkhole and been trapped in the cave, unable to escape, until he died of starvation. They named him Altamura Man, after the nearby town, and began the difficult process of analyzing the fossilized skeleton embedded in the limestone.

Because the skeleton could not be removed without destroying it, scientists at the time were only able to make observations. Still, they discovered that Altamura Man was incredibly important. He had the most complete skeleton of any ancient hominin ever found. Almost every bone in his body had been preserved, including the delicate bones inside his eye sockets. His features were a mix of *Homo neanderthalensis* and other species, which seemed to make him a key source for evidence about

HOMO NEANDERTHALENSIS

Modern humans still carry bits of DNA from ancient hominin species. One of these ancestors is *Homo neanderthalensis*. This species' scientific name means Neander Valley Human, named for the place where their remains were first discovered in Germany in 1856. Often known as Neanderthals, the species lived between 400,000 and 40,000 years ago, alongside early *Homo sapiens*.[3] *Homo neanderthalensis* settled across central Asia and Europe, adapting as they migrated into colder northern climates. To preserve heat, they had short legs and a solid, stocky build. Despite the common portrayal of dull cavemen, *Homo neanderthalensis* were intelligent, skilled toolmakers who spoke some form of language. *Homo sapiens* interbred with *Homo neanderthalensis*, and to this day many modern humans carry a small percentage of Neanderthal DNA.

Approximately 1.5–2.1 percent of the DNA of modern humans from populations outside of Africa comes from *Homo neanderthalensis*.[5]

hominin ancestors. If Altamura Man could be properly studied, he might give scientists new insight into human evolution. But it took nearly 20 years for that to happen.

PIECE BY PIECE

In 2009, researchers began the process of recovering a piece of Altamura Man's bone for analysis. While most of the skeleton was covered in a thick layer of mineral growth, preserved but impossible to access, a few small portions of bone had fractured and fallen away. Using a sterile method adapted from surgical procedures to avoid contaminating the evidence, scientists retrieved these fragments from behind the skeleton.

The extracted bone came from Altamura Man's shoulder blade, which had only a thin layer of mineral residue on it. Scientists also extracted a small piece of broken stalagmite for comparison testing. First, the samples were documented with photographs and X-rays. Then they were cleaned, encased in resin, and thinly sliced for analysis under a microscope. The sections with the most promising samples were sent to laboratories that could determine the age of the mineral residue. Measuring the amounts of particular elements in the mineral residue allowed researchers to determine how long ago it had formed. They concluded that Altamura Man had been in Lamalunga Cave for at least 130,000 years.[4]

Scientists were still not sure which species Altamura Man belonged to, however. The difficulty of accessing the shoulder blade bones without damaging the skeleton postponed further analysis for several years. In 2015, researchers were finally able to extract the remaining fragments of shoulder bone from Lamalunga Cave. After 3D scanning and piecing together the bones from 2009 with the bones from 2015, researchers were able to virtually reconstruct the skeleton's right shoulder blade. They compared the virtual 3D model with other examples of ancient hominin shoulder blades, revealing that Altamura Man was definitely a *Homo neanderthalensis*.

Perhaps the most exciting discovery scientists have made, however, was that Altamura Man's fossilized bones still contained preserved DNA. This DNA is the

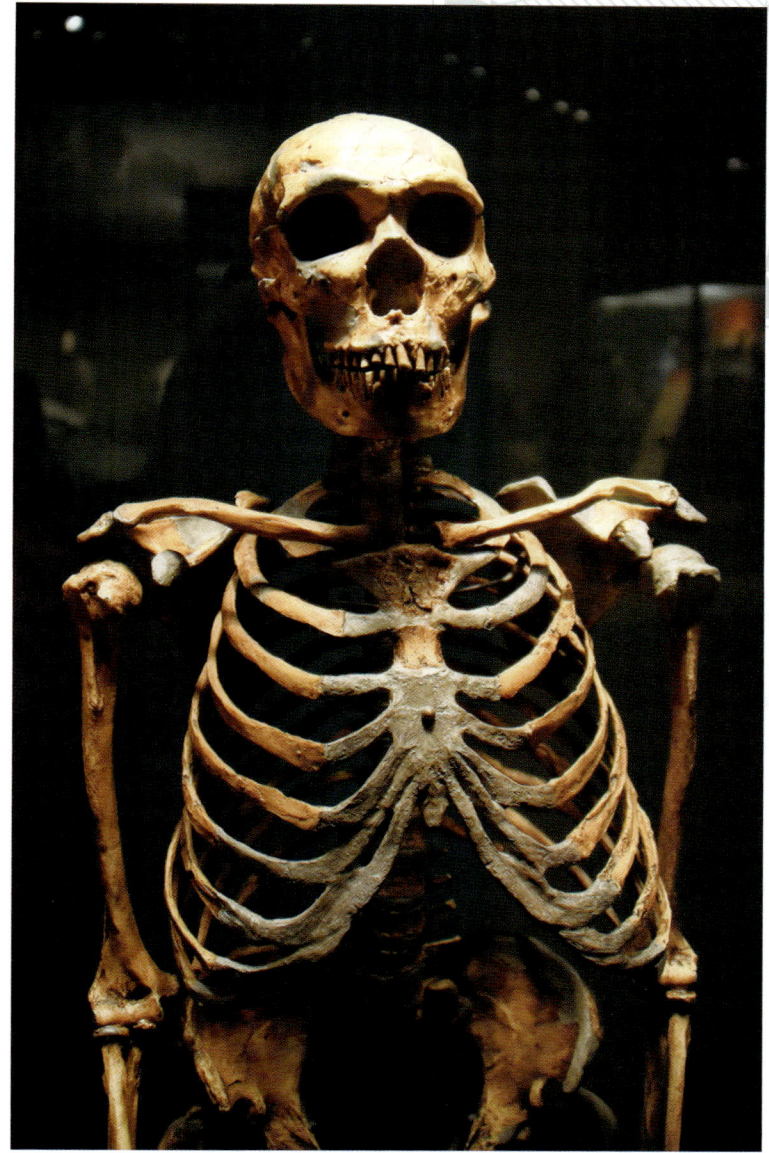

Comparing samples from Altamura Man to known skeletons of *Homo neanderthalensis* helped researchers definitively identify which species the bone came from.

Homo neanderthalensis carved jewelry out of bone and created colored pigments to paint shapes on cave walls.

VOLCANO DISASTER

Forty thousand years ago, the largest European volcanic eruption in hundreds of thousands of years caused a disaster. Scientists have hypothesized that the supereruption might have contributed to the decline and extinction of *Homo neanderthalensis*.

In the area that is contemporary southern Italy, the Campi Flegrei volcano erupted on a massive scale. It spewed ash clouds over an area of 1.4 million square miles (3.6 million sq km). The clouds contained high amounts of sulfur dioxide, causing air pollution as well as widespread temperature changes as they blocked sunlight. For several years, the world cooled by as much as 3.6 degrees Fahrenheit (2°C).[6] This occurred during the last Ice Age, when hominins were already struggling to survive. The effects may have caused severe environmental problems, leading to the decline of *Homo neanderthalensis*.

oldest of any *Homo neanderthalensis* ever found. As of 2019, the DNA was too fragmented to be tested, but researchers expected that in the near future technology would be up to the challenge.

COMING SOON

Altamura Man promises to reveal much more about the hominin ancestors of modern humans. As technologies improve, further analysis of this incredibly well-preserved skeleton will allow scientists to determine how *Homo neanderthalensis* lived, evolved, and passed on elements of its genetic code. Underneath its mineral growths, the skeleton could hold significant clues to the evolutionary path of modern humans.

Because scientists have not been able to excavate Lamalunga Cave and recover Altamura Man, they have studied it mainly through remote video feed. It is often better to preserve a site than to attempt invasive procedures that might damage the evidence forever. Paleoanthropology and paleontology have come a long way from the early days of amateur fossil hunting. Today, scientists in these fields prioritize the preservation of finds for future scientific analysis. This is especially important to prevent errors in the interpretation of a fossil.

THE PILTDOWN MAN HOAX

The traditional idea in paleoanthropology was that there must have been a single "missing link" ancestor between modern humans and ancient apes. In 1912, an amateur English archaeologist claimed to have discovered this key ancestor. He—or an unknown person familiar with the scientific evidence and expectations—staged one of the biggest hoaxes in the history of paleoanthropology, setting back scientific understanding for decades to come.

The fossilized skull he claimed to have found was called Piltdown Man, and experts guessed that it might have been 500,000 years old. In reality, the fossil was a combination of a human skull and an orangutan jaw, stained with dye to match each other and appear the same age. Only a few scientists at the time were allowed to examine the skull, but they claimed it was authentic. The discovery became worldwide front-page news. It wasn't until decades later that the find was completely discredited. By then, the misunderstanding it caused about human evolution was widespread.

INTO THE PIT OF BONES

The rift in the southern slope of the rolling Atapuerca foothills stood out dark against the faded green grass. Paleontology student Trinidad Torres approached the cave entrance, prepared for a trial expedition of experimental fossil hunting. Entering the dim, cool cavern—the portal to a sprawling underground system of tunnels—his eyes adjusted to the light, and he could see the team's excavation trenches. Leaving the work area behind, Torres made his way deeper, moving through a tight passage that squeezed him into a huge cavern with a soaring ceiling.

Here, four passageways came together at a junction point. Torres turned left into a long, narrow passageway. As he hiked down the dark tunnel, his light moved across the walls, revealing first a glimpse of an ancient cave painting, then a section of a shape engraved in the limestone. The long gallery finally came to an end in a constriction point, where Torres was

In the years since Trinidad Torres's expedition, the Atapuerca area has become a bustling archaeological site. ▷

forced to crawl on his hands and knees through to the next cavern, the Chamber of the Cyclops. The chamber was where he had been digging test pits to search for fossilized bones from the ancient cave bears he studied. Now he planned to push on farther into the cave.

A sharp hill sloped upward from the cave floor, and Torres climbed the pile of sediment carefully, emerging at the top on a balcony-like ledge. Before him, a black hole plunged abruptly downward. Slowly, Torres wedged himself into the hole, located sturdy footholds and handholds, and began to descend into the dark. The 43-foot (13 m) shaft would be a dangerous fall, and even when he reached the bottom, Torres' feet touched down on a slanted surface. The chamber, nearly 100 feet (30 m) underground, ramped steeply downward before it leveled off.[1] Torres hiked to the bottom and scanned the cave floor for a good fossil sampling site. A piece of bone caught his eye. It was a jawbone that clearly did not belong to a cave bear. It belonged to a hominin.

CAVE PAINTINGS

Symbolic images painted or engraved on cave walls are some of the most fascinating and mysterious discoveries in paleoanthropology. The images indicate a turning point in human evolution toward abstract thinking. Scientists have believed that the ability to create symbolic art is inherently human, confined to *Homo sapiens*. However, recent evidence from several caves in Spain suggests that *Homo neanderthalensis* were also cave artists. Hominins painted on cave walls more than 65,000 years ago, before modern humans lived in the region. Even earlier, around 115,000 years ago, hominins created colored pigments.[2] It's possible that modern humans were not the only species capable of abstract thinking and the use of symbols.

THE CAVE AND THE PIT

The caves of the Atapuerca Mountains in northern Spain were well known. But it wasn't until 1976, when Trinidad Torres found a hominin fossil there, that scientists realized how significant the site could be. Sima de los Huesos, meaning Pit of Bones, had been explored but never excavated.

The large limestone cave system 12 miles (20 km) outside the town of Burgos was discovered in the late 1890s when railroad workers cutting a trench through the mountains hit a series of caverns. In the 1910s and 1920s, researchers studied a mysterious cave painting of a red horse's head in El Portalón, the main entry chamber to the cave system. By the 1980s, the entry chamber had been fully excavated, confirming it to be a Bronze Age occupation site. A separate chamber nearby with fossilized skeletons, stone tools, and pottery proved to be a burial tomb.

Sima de los Huesos, far from the main cave entrance, was known to cavers but not to researchers. When Torres reported his discovery of a hominin fossil to his university adviser, the find launched a decades-long series of excavations. By 1995, a team

MILLION-YEAR-OLD MAN

The latest big discovery in the Atapuerca Mountains cave system comes from a place near Sima de los Huesos. Part of the original trench excavated by railroad workers more than 100 years ago, Gran Dolina is a cave where extremely old fossilized bones have been found. In 2018, a tooth from the site was analyzed using two different dating methods and discovered to be nearly one million years old. This finding confirmed a theory that the fossilized tooth came from the oldest hominin species in Europe, which is currently unknown but referred to by the placeholder name *Homo antecessor*, meaning Pioneer Human.

led by paleoanthropologist Juan Luis Arsuaga had recovered more than 1,500 hominin fossils from the chamber, as well as hundreds of fossils from cave bears, foxes, and other animals.[3] By studying the layers and arrangement of fossils deposited over time, the team concluded that the pit had served as a burial place for a group of hominins. Later on it became an accidental trap for cave-dwelling animals.

DUELING DATES

In the laboratory, the fossilized bones from Sima de los Huesos were dated relative to a mineral deposit found in the layer above them. A broken limestone stalagmite, analyzed for particular elements, was determined to be 600,000 years old. Assuming the bones were deposited in the pit before the stalagmite was, Arsuaga's team believed the bones to be at least that old. Based in part on that theory, they

Researchers enter the Pit of Bones as part of a 2015 expedition.

HOMO HEIDELBERGENSIS

Little is known for certain about *Homo heidelbergensis*, but many scientists believe it could be the last common ancestor of *Homo sapiens* and *Homo neanderthalensis*. First discovered in Germany in 1907, the species was named for the nearby town of Heidelberg. The Heidelberg Human was tall and strong, with a large brain for its time, but its facial features are similar to earlier species that lacked a robust jaw and chin. These conflicting features make it difficult to match with other fossil specimens or clearly identify the position of *Homo heidelbergensis* in human evolution. The species will likely remain mysterious until researchers are able to use future technology to sequence its DNA.

As recently as 40,000 years ago, there were at least four different hominin species coexisting together in Europe.[5] The only species that survived was *Homo sapiens*.

identified the hominins from Sima de los Huesos as *Homo heidelbergensis*.

In 2012, however, paleoanthropologist Chris Stringer argued that the team had made an error. Studying their data, he concluded the hominin fossils were less than 400,000 years old because they strongly resembled other specimens of *Homo neanderthalensis*. "The skulls, jaws, teeth and skeletons of the Sima fossils show many Neanderthal features," Stringer pointed out, but Arsuaga was not convinced.[4] He agreed that the dating could be incorrect, but he did not think shared features could make a definitive species identification.

After years of debate, new advances in DNA analysis were finally able to settle the disagreement in 2016. By this time, excavators had recovered nearly 7,000 fossils from the pit, representing an astounding site with at least 28 separate individuals. Using newly recovered fossils excavated in sterile conditions, lab researchers sequenced enough DNA to reveal that the hominins in Sima de los Huesos were early relatives of *Homo neanderthalensis* from 430,000 years ago.[6]

EVOLVING ON

The excavation of Sima de los Huesos ultimately challenged and changed scientists' theories about the evolution of hominins. For a long time, scientists assumed that human ancestors evolved in simple and discrete steps, with each successive species being more advanced than the one before. In fact, hominins evolved in what is known as an accretion model. A mosaic of modern traits developed at different times rather than a complete set developing together. Several species could coexist, sharing some features but differing significantly on others. The early *Homo neanderthalensis* relatives are an example of this process.

HEARING AND SPEAKING

When did hominins develop the ability to verbally communicate using language? Answering this question is complex and involves both genetics and physical features. Because verbal communication is closely related to hearing range, evidence of that range can help determine the timeline of language development. Studying fossilized ear bones is the best way to determine when hominins became capable of hearing a wide range of sound frequencies. By creating virtual reconstructions, researchers have learned that ear bones from Sima de los Huesos are similar to ear bones from modern humans in terms of hearing ability. This finding implies that ancient hominins likely used verbal language, though it might not have been a complex language.

Debate is essential for the scientific community, and disagreements often lead to a more complete and complex analysis of discoveries. However, competing theories can be difficult to reconcile, particularly when data is not conclusive. The more puzzling a discovery is, the more valuable it becomes to develop testable theories. Then scientists can perform experiments and research to settle any scientific questions.

SCIENCE CONNECTION

DATING ROCKS AND BONES

Determining the age of material found during an excavation is a rigorous science. Paleontologists and paleoanthropologists commonly rely on radiometric dating to learn the precise ages of their finds. Radiometric dating is a general term for various techniques that measure the amount of natural radiation present in a sample. Based on this measurement, researchers can calculate a date range for the material.

Finds are dated in specialized laboratories, where researchers take a sample and then use a chosen analysis technique to measure the concentration of particular elements in the material. The technique used to date a sample depends on the type of material and estimated age of the find.

Fossils, minerals, and sediments that are less than 500,000 years old can be analyzed with uranium-series dating, which measures concentrations of the elements uranium and thorium in a sample. Because uranium decays into thorium at a steady and specific rate, researchers can use its concentration to calculate how long it has been decaying and determine how old the material is. A similar technique, carbon dating, is useful for younger fossils. Older fossils, up to three million years old, can be analyzed with a technique called electron spin resonance. This method measures the amount of natural radiation that has been absorbed by the material in a sample.

CHAPTER SIX

UNDER THE SERPENT ISLAND

In the shadow of lush tropical mountains, down a narrow, potholed road that often washed out during the rainy season, a team of scientists in a cramped bus approached an ancient cave. Its huge entrance almost seemed to open like a mouth before them, stalactites hanging down like sharp teeth, the lip along the bottom guarding the dim interior. Venturing inside the damp cave, the group gazed up at the soaring ceiling, curved in the shape of a cathedral dome. This was the remote and astonishing site they planned to excavate throughout the next several dry seasons.

As the team neared the end of the third season, however, only a few scientists remained at the cave. Emanuel Wahyu Saptomo oversaw the

Teams of scientists traveled to a remote cave in Indonesia in search of fossils.

Liang Bua means "cool cave" in the local Manggarai language spoken on Flores.

group for the last few weeks of the dig. Below him, local excavators worked nearly 20 feet (6 m) down from the cave floor surface.[1] In the bottom of a treacherous muddy trench, Benyamin Tarus scraped back layers of soil. He began to make out an object emerging. Tarus called out to Wahyu, who climbed down the bamboo ladder into the deep hole as quickly as he could. Together, the pair carefully brushed away wet sediment to reveal a white circle of bone that appeared to be extremely fragile. As the circle grew larger and their discovery came to light, Wahyu realized what they had found: the top of a very small human skull.

BENEATH THE CAVE

The island of Flores, which is part of Indonesia, lies near the island of Bali at the end of an archipelago in the Indian Ocean. Shaped like a snake, Flores is a long, meandering strip of land that spreads from tropical mountain jungles to flooded rice paddies to volcanic lakes. In the foothills of the western mountain ranges, not far from a large market town, is the cave known as Liang Bua.

An Indonesian archaeologist inspects a stone found in the Liang Bua cave.

More than 80 percent of the bones excavated in Liang Bua actually come from rats—including a species of cat-sized giant rat that still lives on Flores.[3]

In 2003, two years into their excavation of Liang Bua, a team of Indonesian and Australian scientists discovered a collection of fossilized bones that shocked and divided paleoanthropologists around the world. At first glance, the skull uncovered in Liang Bua appeared to be that of a young child. But as the excavation continued and the jawbone was exposed, the team realized that its permanent teeth meant that the skull belonged to a very small-statured but fully-grown adult.

A few weeks later, when the dry season ended, excavators had managed to recover most of the remaining bones associated with the skeleton. All of them matched the skull in scale. The team was astounded and perplexed by their discovery. The bones did not match any known hominin species. The team members suspected they could represent a new species, but perhaps they were simply a rare example of a hominin with dwarfism. The mysterious skeleton would need to be reconstructed and analyzed before they could determine the answer.

AN ANCESTRAL PUZZLE

In Australia, an anthropologist with a specialty in skull anatomy studied the Liang Bua fossils and reached a conclusion. One year after the astonishing find, the team released that conclusion in an announcement that stunned the scientific community. The Liang Bua skeleton represented a previously unknown species, which they named *Homo floresiensis*—Flores Human.

The statement caused a controversy related to the skeleton's small stature. Many scientists were unconvinced that an entire hominin species could have adapted so drastically during the estimated time period. *Homo floresiensis* had been tentatively dated to 18,000 years ago based on geological analysis of the trench where it was discovered.[4] However, doubtful scientists argued that by that time *Homo sapiens* were living on Flores, which meant the skeleton was more likely to belong to a human with dwarfism.

To obtain a more accurate date, the team devoted years to close analysis of the cave's stratigraphy. They studied the layers of geology and how they had interacted over time. Liang Bua was a complex site where millennia of tectonic shifting and erosion had displaced particular layers of limestone, clay, and volcanic ash. It was the volcanic evidence that finally provided the key to dating *Homo floresiensis*. Several unbroken layers of tuff, a rock formed out of compacted ash, covered a soil layer that matched the original trench.

ACROSS THE WAVES

As far as scientists have determined, Flores has never been connected to the rest of Indonesia via land bridges, even during the Ice Age when low sea levels exposed migration paths. This fact makes it difficult to theorize how and when hominins might have migrated to Flores. The current theory is that after ancestors of *Homo erectus* traversed the land bridge to Java, they might have settled on nearby Sulawesi. From that island, migration would have been possible by sea, as local ocean currents travel south toward Flores. Evidence exists that ancient hominins occupied Flores as early as one million years ago. This hypothesis continues to develop.

ISLAND ADAPTATIONS

Island populations often show unique adaptations compared to areas where populations are able to roam and migrate freely. The evolution of a species can be highly specific to its location's conditions. In remote or isolated regions with limited population mixing, adaptations are more likely to become extreme. When Charles Darwin visited the Galápagos Islands, the species he observed clarified his theory about the process of evolution. The islands off the coast of Ecuador were an isolated ecosystem, and as a result, they contained bird species with unusual adaptations, including traveling underwater, using tools, and even drinking blood.

The preserved geology completely changed the team's understanding of the timeline of hominin occupation in Liang Bua. Dating the layers of sediment revealed that *Homo floresiensis* could not have inhabited the cave fewer than 60,000 years ago and might have lived as many as 100,000 years ago.[5] This knowledge prompted the team to pursue a more precise date through bone analysis.

BONES AND SPIT

Testing the fragile *Homo floresiensis* bones risked damaging the fossils, but given the high scientific stakes, the team decided that obtaining an accurate date was necessary. The team sent three arm bones from separate trenches around the cave to a laboratory for testing. Using uranium-series analysis to date the finds, researchers measured the concentration of several elements in the fossils. The tests indicated that the arm bones ranged from 66,000 to 87,000 years old—too old to be *Homo sapiens*.[6] More than ten years after the discovery, the team was finally certain of its conclusion: *Homo floresiensis* was a new species of ancient hominin.

A *Homo floresiensis* skull, *left*, is seen next to a modern human skull.

> "It's kind of like Flores is its own little laboratory of early human evolution."[8]
> —Rick Potts, paleoanthropologist

INSIDE LIANG BUA

In 2010, the Smithsonian Institution began to collaborate with Indonesian and Australian institutions to create a 3D model of Liang Bua. Researchers recorded the cave's dimensions using sensitive digital scanners to pick up tiny differences in the rock formations' scale and texture. The result is a stunningly detailed rendering of the interior of Liang Bua, including excavation trenches and pits, which both scientists and students can use to remotely study its archaeology and prehistory.

In 2018, results of a related study were announced. A local group of people, the Rampasasa, had been suggested as genetically related to *Homo floresiensis* due to their short stature. Although the Rampasasa were much taller—just under 5 feet (146 cm) on average, compared to the 3.5-foot (110 cm) skeleton—researchers investigated the theory.[7] The Rampasasa agreed to test their DNA using saliva samples. Since it wasn't possible to extract DNA from the Liang Bua fossils, researchers compared the samples to a DNA database of other modern groups.

The comparison revealed that the Rampasasa had similar genetic ancestry to other populations across Asia, and they had no relationship with *Homo floresiensis*. One fascinating insight scientists gained from the study was that twice on Flores, small stature had evolved independently—once in ancient hominins and once in modern humans.

Scientists estimate that ancient hominins on Flores developed an adaptation for small stature within 300,000 years of their migration. The reasons for this adaptation remain unclear, and researchers continue to study similar cases. At Liang Bua, the team of scientists expects that excavation of the cave will continue for the rest of their careers. There may be much more to discover, including evidence regarding the causes of height adaptation. Island excavations can uncover unforeseen and essential information about the evolution of species.

Scientists search carefully through layers of earth to find fossils, date them, and place them into a wider context.

CHAPTER SEVEN

OVER THE OCEAN OF ICE

In the belly of a heaving ship plowing through the roughest waters on Earth, a team of paleontologists clung to their stomachs. For five days, seasickness had made their voyage miserable. Even though the team members expected the ocean passage to be challenging, most of them were so sick that they spent the journey in their bunks. By the time huge icebergs replaced the choppy waves and land was sighted, the team was eager to begin their real expedition.

Whirring helicopters waited for them on the ship's launch pad, rotors beating through the frigid air. The team members loaded up their gear, tents, and tools into the helicopters. Then they were off, rising slowly and

Expeditions to Antarctica involve ships, helicopters, and some of the harshest conditions on the planet.

hovering above the ship, which was anchored offshore among the vast sheets of sea ice. Turning toward land, the helicopters shuttled the expedition over the white and gray seascape, finally touching down on a small island.

The scientists had arrived in one of the harshest environments in the world. The chill wind swept up waves on the sea channel that flowed around the island. Scouting out a spot for their home base, the team surveyed the barren land, unpacked their tents, and set up camp on the extreme edge of the Antarctic continent.

PREHISTORIC ANTARCTICA

The prehistoric environment of the Antarctic Peninsula was much more temperate than it is today. It was then attached to South America, and its climate was similar to many parts of that continent. Antarctica was a forest of warm-weather ferns, flowering plants, and evergreen trees. Amphibians and water birds gathered near its fish-filled lakes, and marine reptiles swam in its shallow seas. Like Australia, attached on its other side, Antarctica was home to a range of ancient marsupials, along with dinosaur species and other reptiles.

UNDERNEATH THE WORLD

Since the 1990s, the arrival of summer in Antarctica has been followed by expeditions. Groups of American and Argentine researchers, cooperating internationally with other scientists, began making journeys to a cluster of islands off the Antarctic Peninsula. When the sea ice had receded far enough for a research vessel to sail into the narrow channel, paleontologists came from around the world, hoping to gather evidence of prehistoric life on the remote continent.

The first excavation season on the islands lasted only a few months, but researchers recovered a fascinating array of fossils. Curious about the evolution of vertebrate animals, paleontologists wondered whether Antarctica might contain evidence about the end of the era of dinosaurs and the beginning of the era of mammals. During that age of transition, from the late Cretaceous period into the Paleogene period, Antarctica's climate was temperate and the continent was just beginning to separate from Australia and South America.

The evolution, migration, and extinction of species in the warm and forested Antarctica that existed between 100 million and 40 million years ago provided clues to paleontologists about a mysterious event. For much of the period, dinosaurs and reptiles coexisted with mammals, but approximately 66 million years ago, the impact of a giant meteor changed life on Earth very suddenly.[1] The dinosaurs soon died out. What happened in Antarctica around the time of this mass extinction?

SURVIVORS OF DISASTER

One of the first fossilized skeletons discovered in the Antarctic islands, excavated by an Argentine expedition in 1992, was a prehistoric bird. A geologist was working on Vega Island, nearly 40 miles (60 km) north of Argentina's research base, when he came across a curious feature in the rocks he was studying.[2] Protruding slightly from the rock were several small, hollow bones. They seemed to belong to some type of ancient animal capable of flight.

Once the expedition returned to Argentina, the fossilized bones were painstakingly exposed from the rock material. The researchers sent the fossil to a paleontologist in the United States for analysis. Julia Clarke, who specialized in the early evolution of birds, used X-rays to scan the fossil in high resolution. She then created a 3D digital model from the scans. The digital model allowed her to closely examine the fossilized skeleton inside the rock without physically damaging the specimen.

In 2005, Clarke announced her findings. Analysis had determined that the fossil represented a previously unknown species of bird. She named the species *Vegavis iaai*, in honor of its discovery on Vega Island by the Argentine Antarctic Institute, known as IAAI. The announcement excited the scientific community because the prehistoric bird strongly resembled a modern duck, and it had lived

A paleontologist in Antarctica on Vega Island sifts through loose surface material to find small fossils.

Modern birds are living dinosaurs—they are the only known dinosaur lineage to survive the mass extinction event of the Cretaceous period.

THE FIRST BIRD

The earliest known bird is a dinosaur called *Archaeopteryx*, which means "ancient wing." *Archaeopteryx* was discovered in Germany in 1861, shortly after Charles Darwin published his theory of evolution. As a result, the prehistoric bird was considered an ideal "missing link" to illustrate the evolutionary process. Approximately the size of a modern crow, *Archaeopteryx* had several dinosaurian features, including extendable claws, a long bony tail, and sharp-toothed jaws instead of a beak. The bird hunted amphibians and insects, and it could fly only short distances. During the Jurassic period, around 150 million years ago, *Archaeopteryx* lived on an ancient archipelago near Earth's equator.[4]

in Antarctica around 67 million years ago—before the mass extinction event.[3]

Clarke continued to study *Vegavis iaai*. When a new specimen was discovered in 2013, her analysis revealed that the fossil contained preserved soft tissues. The fossilized tissues consisted of the bird's trachea, or windpipe, and its syrinx, or voice box. Comparing the fossilized syrinx to the anatomy of other fossilized prehistoric birds and modern birds, Clarke concluded that *Vegavis iaai* would have produced squawking calls similar to those of modern waterfowl. The presence of this distinctive anatomy in a bird before the mass extinction event provided evidence that some birds survived the event.

THE END OF THE WORLD

In 2005, the American-Argentine research team made another significant discovery on Vega Island. Paleontologists came across a rock formation where high winds had eroded the layers of sediment and exposed a series of vertebrae. Nearby, fossilized shellfish revealed that the site had once been underwater. When the team began to excavate, they found a nearly complete juvenile plesiosaur skeleton.

The five-foot (1.5 m) fossil was challenging to recover. In extreme cold, the team spent weeks excavating gravel, while winds up to 70 miles per hour (110 kmh) flung it back into their trenches.[5] Once they managed to remove the fossil from the frozen ground, transportation presented several problems. While attempting to mix plaster to encase and protect the skeleton, freezing temperatures turned the water to slush. After the team managed to plaster and wrap the fossil in canvas, it proved too heavy to carry on foot. Ultimately, the plesiosaur was airlifted and flown by helicopter to the team campsite.

After the expedition, paleontologists cut away the fossil's wrappings and began to analyze it in their laboratory. The sediments surrounding the plesiosaur proved to be a perfect case study in stratigraphy. Layers of volcanic ash and marine sand indicated that the shallow sea was shaken by a violent eruption and then buried in burning volcanic debris. The plesiosaur, along with several other juvenile marine reptiles, had died instantly and remained preserved in ash for millions of years.

THE MONSTER OF LOCH NESS

In the 1930s, several people in northern Scotland began to claim they had sighted a sea monster long rumored to inhabit a local lake. This lake, called Loch Ness, suddenly became infamous around the world for supposedly harboring an ancient beast. A faked photograph of the animal circulated in newspapers, leading to an intriguing theory: the sea monster was a plesiosaur that had survived extinction by living in the lake since prehistoric times. Even though the idea was scientifically impossible, many tourists visited Loch Ness in hopes of finding evidence. In 1994, it was revealed that the photographer behind the most famous image of the monster had faked his photo.

Fossils of young animals are unusual because their bodies have rarely been preserved. When they died, it was often because a predator ate them.

To paleontologists, the Vega Island site appeared to have been a marine reptile nursery, where prehistoric species raised their young in the shallows of the ancient Southern Ocean. In this warm climate, carnivorous plesiosaurs could swim the open sea hunting fish and krill, their long sharp teeth stabbing or trapping their prey. The plesiosaur's elongated neck and small head aided its ambush technique, while strong diamond-shaped paddles propelled it through the water.

Until 66 million years ago, marine reptiles occupied the top of the ocean food chain. After the apocalyptic meteor impact that abruptly changed Earth's climate, however, these species disappeared. Shortly before that mass extinction event, around 70 million years ago, the juvenile plesiosaur on Vega Island was buried alive by a volcanic eruption. Its fossilized skeleton represents one of the best-preserved examples of vanished prehistoric ocean life.[6]

Scientists continue to search for the reasons why certain Antarctic species went extinct while others survived. From prehistoric dinosaurs to ancient hominins, all species on Earth either evolved and passed on their genetic traits or died out. Recent technological developments have allowed scientists to analyze DNA in much greater detail and in some cases connect modern species with their genetic ancestors.

Plesiosaurs appear in the fossil record as early as 200 million years ago.

UP THE SIBERIAN MOUNTAIN

The cave felt magical from David Reich's first step inside. The ground beneath his feet had preserved vital information for millennia. To understand this astonishing spot, Reich had labored for years—and traveled more than 5,000 miles (8,000 km) around the world.[1]

Over the last few days, Reich had flown from the United States to London, England; through Moscow, Russia; and into Siberia. He had driven ten hours along rutted and winding mountain roads to a remote archaeological site near the Kazakhstan border. There, resting between two mountain peaks in a canyon, lay a small settlement of cabins.

Siberia's Denisova Cave holds evidence of the hominins who dwelled inside it tens of thousands of years ago.

Researchers take great care inside the cave to avoid disrupting or damaging the fragile fossils and artifacts.

Above the cabins, set into a craggy rock face, a tall rectangular opening was almost hidden by summertime foliage. Flights of makeshift wooden stairs led into the opening, where wooden scaffolding formed walkways and strengthened the walls of deep excavation trenches. The high limestone ceiling bloomed with moss and lichen. The trenches stretched downward, marked every few inches with an orange number tag. Here, Reich learned, was where important fossilized information was waiting to be discovered and decoded.

A HIDDEN HOME

Denisova Cave sits 90 feet (28 m) above the Anui River, which flows through the Altai Mountains of central Russia. The river valley was attractive to ancient hominins, who made their homes in the cave for more than 250,000 years.[2] The site provided access to fresh water, an excellent defensive view, and a ceiling that sloped up into a natural chimney, venting the smoke from hearth fires.

Even better, the canyon acted as a funnel for migrating animals, allowing hunters to easily pick off game. Local deer, elk, and bison were plentiful, and the mild climate of the period provided many edible plants, such as acorns and walnuts. Remote as the cave might be today, for thousands of years it was an ideal site. Even in the 1700s, it was home to a hermit known as Denis, who gave the cave its name.

In the late 1970s, a Russian paleontologist visited the cave and excavated a test pit, where he found a cache of ancient stone tools. Within a few years, archaeologists had organized regular expeditions, and by the 1990s a permanent research camp was built. The Siberian archaeological branch of the Russian Academy of Sciences led yearly excavations and collaborated with international researchers to analyze their amazing finds.

IN THE TRENCHES

From the beginning, excavating fossils in Denisova Cave presented several difficulties. The majority of finds came in the form of fragments, such as bits of bone gnawed to pieces by animals, tiny rodent teeth, and chunks of charcoal. Archaeologists developed a careful sorting method to prevent the loss of any potential data.

Inside the cave, excavators hooked each bucket of sediment onto a moving cable that traveled down across the river. Workers emptied the sediment into a mesh basket, sieved it in the river to wash away soil, and sorted it by size using a series of screens. Researchers scrutinized the tiniest sediments, picking out bones and other finds with tweezers. Among the larger finds, researchers discovered tools and jewelry made of stone and animal bone by *Homo sapiens*.

Hominin bones, the ultimate discovery, were rare. Only a handful of the fossilized bones had been collected over the years. The cool, dry cave had preserved them well. A toe bone belonged to *Homo neanderthalensis*. A suspiciously large molar was originally thought to belong to a cave bear. But when the tooth was sent to geneticists for testing it proved, along with a key piece of finger bone, to belong to a previously unknown species of hominin.

DOWN TO THE BONE

In 2010, paleogeneticists in Germany identified two small bones from Denisova Cave that did not fit any known genetic pattern of ancient hominins. For thousands of years, this

HOMO DENISOVA

Though they have studied its DNA, scientists know remarkably little about the physical characteristics of *Homo denisova*. The only known fossils of *Homo denisova* are a small collection of teeth and bone fragments that give little indication of the species' size, features, or abilities. Instead, the fossils have produced a trove of genetic information that is rare even for well-preserved hominin bones. Scientists have learned that *Homo denisova* shared a common ancestor with *Homo neanderthalensis*, migrated across central Asia 500,000 years ago, and passed on genes to *Homo sapiens* in Southeast Asia before its eventual extinction.[4]

mysterious species had inhabited the cave, leaving behind tiny breadcrumbs of data—their DNA. Paleogeneticist Svante Pääbo, collaborating with Reich's laboratory in the United States, used DNA samples and cutting-edge technology to sequence the species' complete set of genes. The finished genome turned out to be revelatory.

Archaeologists had determined that *Homo denisova*—named after the cave—had at times occupied the same site as *Homo neanderthalensis* and *Homo sapiens*. This insight was enhanced by the results from comparing the species' genomes. *Homo denisova* was distantly related to *Homo neanderthalensis*. At some point in ancient history, the two lineages had split, but evidently they still shared some territory.

Even more groundbreaking, a comparison of *Homo denisova*'s genome to that of modern humans in Southeast Asia revealed several genes inherited from the ancient hominin. Sequencing genomes, a relatively new capability, led to a breakthrough in paleoanthropology. The interconnected relationships of hominin species and early humans became an important area of study.

ANCIENT ANCESTRY

Today, Denisova Cave lies in southern Russia near the border of Kazakhstan, not very far from China and Mongolia. In the ancient past, its location might have been adjacent to various hominin territories. The genetic mixing that occurred in the ancient past is detectable today in modern humans. Geneticists have discovered that particular evolutionarily beneficial genes that still appear in Inner Asian, Southeast Asian, and Oceanian populations were inherited from *Homo denisova*. For instance, geneticists believe that Tibetan people inherited a gene that makes it easier for them to survive at high mountainous altitudes. The early hominin contributed up to five percent of modern aboriginal Australian DNA, and up to six percent of modern New Guinean DNA. In a similar manner, *Homo neanderthalensis* contributed up to two percent of modern Asian and European DNA.

Scientists are able to learn a surprising amount from small fragments of hominin remains, such as molars.

In 2017, researchers launched a project to sequence more genomes from Denisova Cave. Researchers in England had developed a technique to distinguish between fragmentary animal and hominin bones. They used mass spectrometry to examine the type of collagen in a fossilized bone, enabling them to rapidly identify hominin remains. The technology provided geneticists in Germany with several new DNA samples for sequencing and analysis.

In a spectacular stroke of luck, the first hominin bone sent to the laboratory revealed a lost chapter of ancient history. Pääbo's team determined that the fragment of bone belonged to a young woman who had lived in Denisova Cave approximately 100,000 years ago. Her genome showed an unexpected lineage—half of her genes matched *Homo denisova* while the other half matched *Homo neanderthalensis*. Her mother and father had come from two different species that inhabited the region, making the young woman the first generation of a new mixed ancestry.[5]

The discovery further energized the search for ancient DNA. Only a few years earlier, paleoanthropologists had considered such tiny bones untestable. "I would have said we will never find this," Pääbo admitted. "It is like finding a

LAGAR VELHO BOY

The young woman from Denisova Cave is not the only ancient hominin discovery with signs of mixed ancestry. In 1998, the fossilized skeleton of a young boy was found at the Lagar Velho archaeological site in Portugal. The four-year-old child had been buried approximately 25,000 years ago in a ritual that integrated burned pine branches, ornaments of shell and bone, and a red pigment made of ground ochre.[6] The proportions of his skeleton indicated a close relationship with *Homo neanderthalensis*, while his reconstructed skull appeared similar to *Homo sapiens*. Scientists continue to debate whether Lagar Velho Boy displays mixed *Homo neanderthalensis* and *Homo sapiens* ancestry.

needle in a haystack."[7] But the young woman, now known as Denny, revealed a fascinating glimpse of what geneticists hoped to learn about ancient genetic lineage.

At Denisova Cave, excavation will resume each summer for many more years. In the laboratory, the genome sequencing project will continue as long as it receives funding. Each in their particular discipline, scientists are focused on investigating the archaeological record to learn more about human evolution and genetic ancestry. As DNA analysis enters an era of decoding ancient connections, every fossil discovery has the potential to provide revolutionary information about the history of life on Earth.

Pääbo has won major scientific awards for his breakthrough studies of hominin DNA.

SCIENCE CONNECTION

SEQUENCING A GENOME

Analyzing ancient hominin DNA is known as paleogenetics, a discipline that made significant advances in the late 1990s as genetic technology developed rapidly. Early on, it was only possible to extract and analyze mitochondrial DNA—short sequences inherited from an individual's mother. But advancing technology made it feasible to extract and analyze nuclear DNA—a wider set inherited from both parents. Nuclear DNA can be used to sequence an individual's genome, which comprises their complete set of genes. This involves a multi-step process:

Extracting: Researchers grind up ancient bone tissue and dissolve it using chemicals that isolate the DNA strands.

Tagging: Researchers tag the DNA strands with molecules that act as barcodes.

Copying: Researchers heat the DNA strands, which causes each helix to separate. Each half then replicates itself to generate a new helix, producing multiple copies of each DNA strand.

Sequencing: Researchers add chemical color to each base element in the DNA strand sequences so that software can read the sequences.

Authenticating: Researchers discard any remaining incorrect sequences.

Aligning: Researchers sequence the DNA strands into a genome, using software to read and align overlapping DNA strand sequences.

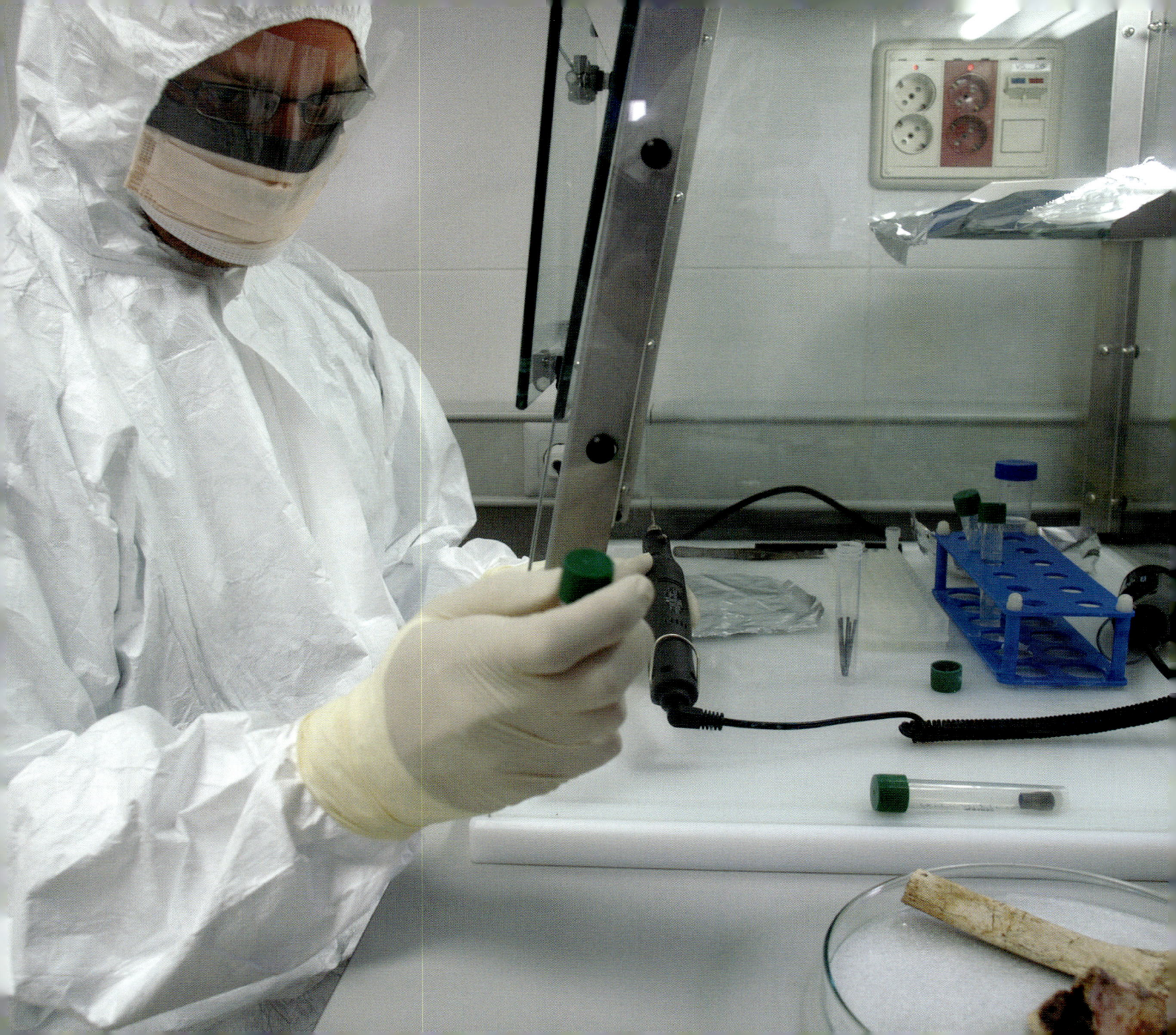

INTO THE FLOODED CAVE

In the still, cool water of the dim tunnel, Alberto Nava scanned his surroundings as he ventured deeper into the maze of underground caves. Two of his friends tracked the guide wire that stretched along the passageway, leading the cave divers and marking their path of return. Eerie blue light from their lamps rippled across the limestone tunnel, picking out the protruding, ancient stalactites and stalagmites. Dodging the hazardous growths, Nava felt his heart rate begin to increase. An hour into the dive, the divers were reaching the end of the line. The guide wire terminated just ahead, and beyond that lay unexplored depths.

Divers explore an underwater cave in Mexico in search of fossils.

The divers swam on, noting their path and gathering data on the caves. Suddenly, exiting a tunnel, darkness wiped out Nava's vision. The light of the lamps was eaten by black nothingness. Nava stopped dead, his heart pumping, his senses utterly disoriented. It was as if he had slipped into a black hole in outer space. Slowly, he cast his light around him, and he realized that he and his friends were floating above a massive pit. Thrilled by their unexpected discovery, the divers cautiously descended into the dark.

More than 100 feet (30 m) down, Nava could almost see the bottom of the pit.[1] The stone seemed to be littered with debris. They were bones, he realized, as his light found surface to illuminate—huge, prehistoric animal bones. Nava and his friends exchanged amazed glances. Only a few feet below them, a long-lost graveyard of extinct species spread as far as their lights could penetrate. Then, as Nava raised his lamp a little higher, he illuminated the ultimate find: an ancient human skull.

UNDERWATER ARCHAEOLOGY

Divers who work with scientists on subaquatic archaeological sites are highly trained. They use scuba gear, a life support system, to access deep sites or underwater structures such as caves where returning to the surface for air is impossible. Strapped to their backs are air supply tanks that allow them to breathe safely at depth. Rebreathers, which recycle the divers' air, increase the amount of time they are able to work underwater. In many ways, the experience of deep diving is comparable to being an astronaut in outer space.

DOWN THE BLACK HOLE

The Yucatán Peninsula in the south of Mexico is home to areas of remote jungle and is a busy tourist region. Underneath it, inside the limestone bedrock, hides an extensive and mostly unexplored system of underwater tunnels and caves. Shaped long ago by erosion, the formations are known as cenotes—pits or natural well holes. Only trained cave divers enter these waters, and these expeditions are dangerous even for them. Many divers have died after losing their way and running out of oxygen. Until advanced diving technology was invented, the cenote system was essentially inaccessible, as it had been for thousands of years.

Many prehistoric sites lie off the coast of the Americas under the ocean, the result of flooding after the last Ice Age.

In 2007, Nava and his friends planned a dive to map a previously uncharted area of the cenote system. Their dive was part of an archaeological project to develop an atlas of the hundreds of miles of tunnels under the Yucatán, which had long been rumored to contain preserved prehistoric human skeletons. During the last Ice Age, before the tunnels were submerged, the cenotes inside the caves were one of the region's few reliable sources of fresh water. Animals and early humans often braved the darkness to access this well water. The ancient well discovered by Nava's team was unusually vast—200 feet (60 m) wide and nearly 150 feet (45 m) deep.[2] They named it Hoyo Negro, meaning "black hole."

UP FROM THE ICE AGE

At the bottom of Hoyo Negro, scattered across the stone floor, Nava's team found an astonishing array of skeletal remains. Extinct species of ancient giant animals made up half of the fossils. A massive pelvis bone and femur represented a gomphothere, an animal similar to the ancient mastodon and the modern elephant. The divers counted more than 30 skeletons of Ice Age creatures, including cave bears, giant sloths, and saber-toothed tigers.[3]

When the divers discovered human remains, however, they recognized that the site had major significance to the study of human history. The skull lay on a ledge, and nearby were more bones—a nearly complete skeleton. Nava's team reported its finds to Pilar Luna, the director of subaquatic archaeology at Mexico's National Institute of Anthropology and History, who coordinated an excavation. In 2014, a team of cave divers and scientists began a carefully rehearsed series of dives to recover the skeleton. Their dives were restricted by the amount of oxygen they carried, the travel time to reach the cenote and safely return, and the risks of transporting fragile bone. Working with a team

Divers carefully collect a human skull found in Hoyo Negro.

of archaeologists on the surface, the divers carefully extracted first the skull and then the associated bones from Hoyo Negro.

Archaeologist Jim Chatters identified the skeleton as an adolescent girl who was approximately 16 years old when she died. The divers named her Naia, after the Greek water nymph. Her bones had been well preserved in the cenote, fossilized by the still, consistently cool water that contained calcifying minerals and very little oxygen. Evidence suggested that the humans and animals that ended up in Hoyo Negro had died swift deaths caused by falling into the deep pit. Then, as sea levels rose 10,000 years ago, the cenote system filled with water.[4]

READING BONES

In the laboratory, archaeologists reconstructed and examined the young girl's skeleton for evidence of her life and death. Remarkably few of her bones were missing. She had the most complete early human skeleton ever discovered in the Americas. The stable environment of the deep cenote had been ideal for conserving the bones' structural integrity. Archaeologists detected evidence of Naia's short and harsh life. Her limbs showed extensive muscle

MASTODON MYSTERY

In 2017, researchers announced that they had found stunning evidence of ancient hominin occupation in California. The evidence was too old to be associated with *Homo sapiens*, so researchers believed it could only represent an earlier hominin species. The team members had analyzed a collection of ancient mastodon bones that showed signs of skilled butchery. They concluded that 130,000 years ago, unknown hominins had used stone tools to open the mastodon bones, perhaps to extract the nutritious marrow.[5] The claim caused controversy among archaeologists because it contradicted decades of research about migration to the Americas. Debate continues, and the issue will likely remain unresolved for some time.

development, indicating a nomadic life of hunting and gathering. Her left forearm had once been fractured, and the proportions of her limbs indicated she had been periodically malnourished and stood only five feet (150 cm) tall. Her broken pelvis occurred at the time of her death, but before that, the bone had been damaged by childbirth.[6]

Researchers tested the enamel protecting Naia's teeth using a radiometric dating technique. They measured the concentration of carbon that remained in the enamel. Because carbon decays into nitrogen at a slow, predictable rate, researchers are able to use its level of concentration in a sample to calculate the age of the material. In a sterile laboratory, a technician sampled Naia's tooth enamel and dissolved it into a solution. By heating the solution and then freezing it, the technician distilled the liquid into powdered carbon. The amount of carbon decay in the sample indicated that Naia had been lying in Hoyo Negro for approximately 13,000 years.[7]

Some of the prehistoric animals in Hoyo Negro are even older than Naia— the elephant-like gomphothere may be up to 36,000 years old.[8]

EVA OF NAHARON

The underwater cave system of the Yucatán Peninsula preserved several fossilized human skeletons, including the oldest known human remains in the Americas. Called Eva of Naharon after the area where she was discovered, the woman was in her twenties when she died approximately 13,600 years ago.[9] Scientists believe that she died elsewhere and was laid to rest by her family in a place they considered a spiritual underworld. Her family would have been hunter-gatherers, and their presence in Central America is mysterious. Scientists are unsure when and how they migrated so far south beyond the ice sheets covering North America.

Finally, researchers were able to extract mitochondrial DNA from Naia's teeth. Mitochondrial DNA contains short segments of genetic code inherited from a person's maternal lineage. Naia's DNA revealed that for thousands of years, her ancestors had inhabited the Ice Age landmass of Beringia, between contemporary Siberia and Alaska. At some point, these ancestors migrated southward, landing near the cenotes of the Yucatán Peninsula. Naia eventually died inside one of these caves, and her remains waited hundreds of generations to be discovered.

Fossil hunters are, in a sense, explorers of an ancient world that no longer exists. Searching for evidence of Earth's past, they might travel around the globe or go for a walk near their homes. But to understand how life has evolved, they must follow where the fossils lead them. "When we go in these caves," Nava reflected on his fossil hunting

Nava, *left*, discussed the researchers' findings at a Mexico City press conference in May 2014.

Fossil hunters around the globe continue to make discoveries that enhance our understanding of our planet's past.

experience, "we think that we're doing this great exploration, but in reality the original explorers of these tunnels were Naia and her people. We're just following them 13,000 years later."[10] By retracing the footsteps of past life, scientists improve our understanding of the vast and complicated history of life on our planet.

HOMO SAPIENS

Approximately 300,000 years ago in Africa, *Homo sapiens* began to evolve from *Homo heidelbergensis*. By 160,000 years ago, these ancient humans had evolved into what scientists consider modern humans. Their skulls were broad at the top and round in the back, and they had a small face and jaw with a protruding nose and chin. Around 100,000 years ago, groups of modern humans began to migrate out of Africa. Humans inhabited Africa, Asia, Europe, Australia, and likely the Americas by 40,000 years ago, when a widespread evolutionary shift occurred. Around that time, humans began to form complex cultures that used verbal and written language, produced clothing and handcrafts, routinely practiced personal adornment, and created art and music. Humans remained migratory hunter-gatherers until at least 11,000 years ago, when groups began to domesticate plants and animals.[11]

ESSENTIAL FACTS

SIGNIFICANT EVENTS

▶ Mary Anning discovered a fossilized marine reptile skeleton in England in 1811.

▶ Researchers uncovered the remains of a newly discovered hominin species, *Homo floresiensis*, in Indonesia in 2003.

▶ Scientists discovered the hominin species *Homo denisova* in Russia in 2010.

▶ The Rising Star Expedition discovered the hominin species *Homo naledi* in South Africa in 2013.

▶ Divers retrieved ancient human fossils from a deep underwater cave in Mexico in 2014.

KEY PLAYERS

▶ Early fossil hunter Mary Anning discovered marine reptiles.

▶ Archaeologist Wahyu Saptomo discovered *Homo floresiensis*.

▶ Paleogeneticist Svante Pääbo sequenced the genome of *Homo denisova*.

▶ Paleoanthropologist Lee Berger discovered *Homo naledi*.

▶ Underwater archaeologist Alberto Nava helped uncover human fossils in the cenotes of Mexico's Yucatán Peninsula.

IMPACT ON SCIENCE

Fossil hunters and their discoveries have changed humanity's understanding of life on Earth, challenged scientific and historical worldviews, and advanced the science of evolution. Studying everything from fearsome dinosaurs to human ancestors, scientists have used modern techniques and technologies to turn preserved bones into vivid windows into the past.

QUOTE

"We have all these things we think of as human. . . . Walking upright is human, a large brain is human. . . . But all of these things happened at different times in different ancestors. The package we think of as human did not appear simultaneously."

—*John Hawks, paleoanthropologist*

GLOSSARY

accretion model
The theory that human traits appeared gradually over hundreds of thousands of years.

adaptation
A change in traits within a population that is caused by different conditions in the environment.

archaeology
The scientific study of human activity through the excavation of sites and analysis of artifacts.

embryo
An animal still developing in the womb.

evolution
Descent from a common ancestor with modification over many generations.

fossil
The physical remains or impressions of a prehistoric organism preserved in some form.

genome
An organism's genetic material.

hominin
A member of a family of mammals that walk on two legs, including humans and human ancestors.

mass spectrometry
A method of identifying the chemical structure of a substance by particle mass and charge.

matrix
The physical material in which archaeological artifacts are located.

paleoanthropology
The scientific study of ancient hominin behavior and culture.

paleontology
The study of past life, involving fossils and previous geological periods.

radiometric dating
Methods of dating fossils that measure the conversion of an unstable element into a stable element.

repatriate
To return something to the country in which it originated.

stalactite
A structure hanging from a cave ceiling, made of calcium salts deposited by dripping water.

stalagmite
A column rising from a cave floor, made of calcium salts deposited by dripping water.

stratigraphy
The scientific study of layers of sediment, soil, and artifacts at an archaeological site.

ADDITIONAL RESOURCES

SELECTED BIBLIOGRAPHY

Boodhoo, Thea. "Saving Mongolia's Dinosaurs and Inspiring the Next Generation of Paleontologists." *EARTH Magazine*, 15 Feb. 2017. earthmagazine.org.

Jaggard, Victoria. "These Are the Dinosaurs That Didn't Die." *National Geographic*, May 2018. nationalgeographic.com.

Liesowska, Anna. "First Glimpse Inside the Siberian Cave That Holds the Key to Man's Origins." *Siberian Times*, 28 July 2015. siberiantimes.com.

NOVA. "First Face of America." *PBS*, 7 Feb. 2018. pbs.org.

Wong, Kate. "Mysterious New Human Species Emerges from Heap of Fossils." *Scientific American*, 10 Sept. 2015. scientificamerican.com.

FURTHER READINGS

Edwards, Sue Bradford. *The Evolution of Mammals*. Abdo, 2019.

Hand, Carol. *The Evolution of Birds*. Abdo, 2019.

Roberts, Alice. *Evolution: The Human Story*. DK, 2018.

ONLINE RESOURCES

To learn more about fossil hunters, please visit **abdobooklinks.com** or scan this QR code. These links are routinely monitored and updated to provide the most current information available.

MORE INFORMATION

For more information on this subject, contact or visit the following organizations:

AMERICAN MUSEUM OF NATURAL HISTORY

Central Park West at Seventy-Ninth St.

New York, NY 10024

212-769-5100

amnh.org

The museum exhibits a global collection of scientific and cultural artifacts, and it supports the research of human cultures, the natural world, and the universe.

DINOSAUR NATIONAL MONUMENT

11625 E 1500 S

Jensen, UT 84035

435-781-7700

nps.gov/dino/index.htm

Dinosaur National Monument is a protected area rich with dinosaur fossils. Visitors to the monument can see fossils and learn about the work of paleontologists.

SMITHSONIAN NATIONAL MUSEUM OF NATURAL HISTORY

Tenth St. at Constitution Ave. NW

Washington, DC 20560

202-633-1000

naturalhistory.si.edu

The museum educates visitors about the natural world and supports researchers by preserving and exhibiting a collection of artifacts documenting the history of the planet and humankind.

SOURCE NOTES

CHAPTER 1. INTO THE DEEP TOMB

1. "Deep in the Dark Zone." *National Geographic*, 2015, blogs.plos.org. Accessed 21 Feb. 2019.

2. Terry Devitt. "Journey to Amazing Hominin Discovery Started on Facebook." *University of Wisconsin–Madison*, 2015, news.wisc.edu. Accessed 21 Feb. 2019.

3. Tabitha M. Powledge. "All About the Fossilized Bones of (Maybe) Homo naledi." *PLOS On Science Blogs*, 11 Sept. 2015, blogs.plos.org. Accessed 21 Feb. 2019.

4. Julien Benoit. "It's Time to Celebrate Africa's Forgotten Fossil Hunters." *Conversation*, 15 May 2017, theconversation.com. Accessed 18 Feb. 2019.

5. Terry Devitt. "Homo Naledi." *University of Wisconsin–Madison*, 2015, news.wisc.edu. Accessed 21 Feb. 2019.

6. Devitt, "Journey to Amazing Hominin Discovery Started on Facebook."

7. Devitt, "Hip Heaven."

8. James de Villiers. "You Can Now Become One of SA's 'Underground Astronauts'—If You Can Squeeze through a 18cm Hole." *Business Insider South Africa*, 6 Sept. 2018, businessinsider.co.za. Accessed 21 Feb. 2019.

9. Ann Gibbons. "World's Oldest Homo Sapiens Fossils Found in Morocco." *Science*, 7 June 2017, sciencemag.org. Accessed 21 Feb. 2019.

10. Paul HGM Dirks et al. "The Age of Homo Naledi and Associated Sediments in the Rising Star Cave, South Africa." *eLife*, 9 May 2017, elifesciences.org. Accessed 21 Feb. 2019.

11. Kate Wong. "Mysterious New Human Species Emerges from Heap of Fossils." *Scientific American*, 10 Sept. 2015, scientificamerican.com. Accessed 18 Feb. 2019.

12. "Species." *National Museum of Natural History*, 21 Feb. 2019, humanorigins.si.edu. Accessed 21 Feb. 2019.

CHAPTER 2. ALONG THE JURASSIC COAST

1. Marie-Claire Eylott. "Mary Anning: the Unsung Hero of Fossil Discovery." *Natural History Museum*, 9 Mar. 2018, nhm.ac.uk. Accessed 18 Feb. 2019.

2. Anthea Lacchia. "558m-year-old Fossils Identified as Oldest Known Animal." *Guardian*, 20 Sept. 2018, theguardian.com. Accessed 28 Feb. 2019.

3. Eylott, "Mary Anning."

4. Doug Stewart. "Mary Anning." *Famous Scientists*, 26 Oct. 2016, famousscientists.org. Accessed 28 Feb. 2019.

5. Henry Neville Hutchinson. *Extinct Monsters*. Chapman & Hall, 1893. 38.

6. Patrick Sawer. "The Fitting Find of Fossil-Hunter Mary Anning's Birthday Token, 200 Years on." *Telegraph*, 11 Nov. 2015, telegraph.co.uk. Accessed 18 Feb. 2019.

CHAPTER 3. UP THE FLAMING CLIFFS

1. Roy Chapman Andrews. *The New Conquest of Central Asia: A Narrative of the Explorations of the Central Asiatic Expeditions in Mongolia and China, 1921–1930.* American Museum of Natural History, 1932. 161.

2. Andrews, *The New Conquest of Central Asia*, 208–211.

3. "The Fighting Dinosaurs." *American Museum of Natural History*, 2000, amnh.org. Accessed 12 Mar. 2019.

4. Thea Boodhoo. "The Flaming Cliffs: History." *Institute for the Study of Mongolian Dinosaurs*, 2016, flamingcliffs.org. Accessed 18 Feb. 2019.

CHAPTER 4. DOWN THE LONG SINKHOLE

1. "Grossa Lamalunga," *Showcaves*, n.d., showcaves.com. Accessed 17 July 2019.

2. Beth Blaxland and Fran Dorey. "The First Migrations out of Africa." *Australian Museum*, 11 Feb. 2018, australianmuseum.net.au. Accessed 17 July 2019.

3. Lisa Hendry. "Who Were the Neanderthals?" *Natural History Museum*, 5 May 2018, nhm.ac.uk. Accessed 17 July 2019.

4. Bob Yirka. "Altamura Man Yields Oldest Neanderthal DNA Sample." *Phys.org*, 3 Apr. 2015, phys.org. Accessed 17 July 2019.

5. Charles Q. Choi. "Oldest Neanderthal DNA Found in Italian Skeleton." *LiveScience*, 10 Apr. 2015, livescience.com. Accessed 6 Mar. 2019.

6. Charles Q. Choi. "Ancient Super-Eruption Larger Than Thought." *LiveScience*, 21 June 2012, livescience.com. Accessed 7 Mar. 2019.

CHAPTER 5. INTO THE PIT OF BONES

1. J. L. Arsuaga, et al. "The Sima de los Huesos and the Cueva Mayor–Cueva del Silo Cave System." *Journal of Human Evolution*, vol. 33, 1997, pp. 109–127.

2. Michael Greshko. "World's Oldest Cave Art Found—And Neanderthals Made It." *National Geographic*, 22 Feb. 2018, news.nationalgeographic.com. Accessed 15 Mar. 2019.

3. J. L. Arsuaga et al, "The Sima de los Huesos and the Cueva Mayor–Cueva del Silo Cave System."

4. Robin McKie. "Scientists Are Accused of Distorting Theory of Human Evolution by Misdating Bones." *Guardian*, 9 June 2012, theguardian.com. Accessed 14 Mar. 2019.

5. Todd Shilton. "Chew On This: Neanderthal Jaws Evolved Before Brains." *Conversation*, 19 June 2014, theconversation.com. Accessed 14 Mar. 2019.

6. Max-Planck-Gesellschaft. "400,000-Year-Old Fossils from Spain Provide Earliest Genetic Evidence of Neanderthals." *ScienceDaily*, 15 Mar. 2016, sciencedaily.com. Accessed 12 Mar. 2019.

SOURCE NOTES CONTINUED

CHAPTER 6. UNDER THE SERPENT ISLAND

1. Hari Kurniawan. "Sutikna Traces Man's Ancestry through Liang Bua Find." *Jakarta Post*, 17 Apr. 2005, thejakartapost.com. Accessed 23 Mar. 2019.

2. "Komodo Dragon." *Encyclopedia Britannica*, 2019, britannica.com. Accessed 24 Mar. 2019.

3. "A Day at Liang Bua." *Paige Madison*, 28 Apr. 2017, fossilhistorypaige.com. Accessed 17 July 2019.

4. Carolyn Gramling. "The 'Hobbit' Was a Separate Species of Human, New Dating Reveals." *Science*, 30 Mar. 2016. sciencemag.org. Accessed 18 Feb. 2019.

5. Gramling, "The 'Hobbit' Was a Separate Species of Human."

6. Gramling, "The 'Hobbit' Was a Separate Species of Human."

7. Michael Westaway and Francis David Bulbeck. "We Know Why Short-Statured People of Flores Became Small—But for the Extinct 'Hobbit' It's Not So Clear." *Conversation*, 2 Aug. 2018, theconversation.com. Accessed 18 Feb. 2019.

8. Maya Wei-Haas. "The 'Hobbit' Lineage May Be Much Older Than Previously Thought." *Smithsonian Magazine*, 8 June 2016, smithsonianmag.com. Accessed 24 Mar. 2019.

CHAPTER 7. OVER THE OCEAN OF ICE

1. Mary Bagley. "Cretaceous Period: Animals, Plants & Extinction Event." *LiveScience*, 7 Jan. 2016, livescience.com. Accessed 26 Mar. 2019.

2. Federico Kukso. "While Dinosaurs Romped, Birdsongs Filled the Air in Balmy Antarctica." *Scientific American*, 22 Oct. 2016, scientificamerican.com. Accessed 27 Mar. 2019.

3. Victoria Jaggard. "These Are the Dinosaurs That Didn't Die." *National Geographic*, May 2018, nationalgeographic.com. Accessed 27 Mar. 2019.

4. Joseph Castro. "Archaeopteryx: The Transitional Fossil." *LiveScience*, 14 Mar. 2018, livescience.com. Accessed 28 Mar. 2019.

5. Heidi Ledford. "Rare Reptile Fossil Found in Antarctica." *Nature*, 12 Dec. 2006, nature.com. Accessed 27 Mar. 2019.

6. "Volcanic Blast Likely Killed and Preserved Juvenile Fossil Plesiosaur Found In Antarctica." *National Science Foundation*, 11 Dec. 2006, nsf.gov. Accessed 18 Feb. 2019.

CHAPTER 8. UP THE SIBERIAN MOUNTAIN

1. Debra Bradley Ruder. "Spelunking for Genes." *Harvard Medicine*, Winter 2012, hms.harvard.edu. Accessed 18 Feb. 2019.

2. Anna Liesowska. "First Glimpse Inside the Siberian Cave That Holds the Key to Man's Origins." *Siberian Times*, 28 Jul. 2015, siberiantimes.com. Accessed 29 Mar. 2019.

3. Liesowska, "First Glimpse Inside the Siberian Cave That Holds the Key to Man's Origins."

4. Carl Zimmer. "In Neanderthal DNA, Signs of a Mysterious Human Migration." *New York Times*, 4 July 2017, nytimes.com. Accessed 18 Feb. 2019.

5. Carl Zimmer. "High Ceilings and a Lovely View: Denisova Cave Was Home to a Lost Branch of Humanity." *New York Times*, 30 Jan. 2019, www.nytimes.com. Accessed 18 Feb. 2019.

6. "Lagar Velho." *Encyclopedia Britannica*, 2015, britannica.com. Accessed 1 Apr. 2019.

7. Robin McKie. "Meet Denny, the Ancient Mixed-Heritage Mystery Girl." *Guardian*, 24 Nov. 2018, theguardian.com. Accessed 18 Feb. 2019.

CHAPTER 9. INTO THE FLOODED CAVE

1. "First Face of America." *PBS*, 7 Feb. 2018, pbs.org. Accessed 2 Apr. 2019.

2. Nora Rappaport. "Giant Underwater Cave Was Hiding Oldest Human Skeleton in the Americas." *National Geographic*, 2 Nov. 2016, blognationalgeographic.org. Accessed 18 Feb. 2019.

3. "Hoyo Negro—Fauna." *University of California, San Diego*, 2018, hoyonegro.ucsd.edu. Accessed 2 Apr. 2019.

4. "Hoyo Negro—Fauna."

5. "Humans in America '115,000 Years Earlier Than Thought.'" *Phys.org*, 26 Apr. 2017, phys.org. Accessed 2 Apr. 2019.

6. "First Face of America."

7. "First Face of America."

8. "Hoyo Negro—Fauna."

9. "Eva of Naharon." *PBS*, 2017, www.pbs.org. Accessed 2 Apr. 2019.

10. Rappaport, "Giant Underwater Cave Was Hiding Oldest Human Skeleton in the Americas."

11. Fran Dorey. "Homo Sapiens—Modern Humans." *Australian Museum*, 12 Nov. 2018, australianmuseum.net.au. Accessed 21 Feb. 2019.

INDEX

Altamura Man, 41–44
American Museum of Natural History, 29, 31, 37
ancient fossil hunters, 6, 28
Andrews, Roy Chapman, 26–31
Anning, Mary, 14–19, 22–25
Antarctica, 66–74
Archaeopteryx, 72
Argentina, 34, 70
Argentine Antarctic Institute, 70
Arsuaga, Juan-Luis, 50–52

Berger, Lee, 7–9
Beringia, 96
birds, 33–34, 69, 70, 72
body fossils, 20
Bolortsetseg Minjin, 34–37

cave paintings, 6, 46, 48, 49
cenotes, 91, 94, 96
Chatters, Jim, 94
China, 26, 34, 81
Clarke, Julia, 70–72
coprolites, 19

Darwin, Charles, 62, 72
dating rocks and bones, 52, 54, 61, 62, 95
Denisova Cave, 76–85
Denny, 85
Dickinsonia, 18
Dimorphodon, 23
Dinaledi Chamber, 6–12
dinosaurs, 6, 28–37, 68, 69, 72
DNA, 41, 42, 43–44, 52, 64, 74, 80, 81–85, 86, 96
drones, 7

eggs, 28, 29, 30–33
England, 14, 23, 76, 83
Eva of Naharon, 95
extinction, 28, 34, 44, 69, 72, 73, 74, 80, 90, 91

Flaming Cliffs, 28, 29, 31, 34, 37
Flores, 58–61, 64
footprints, 6, 20
Fossil Depot, 19, 22
fossil repatriation, 37

gene sequencing, 52, 81, 83, 85, 86
Geological Society, 22–23
Germany, 40, 41, 52, 72, 80, 83
Gobi Desert, 26, 29
Gran Dolina, 49
Gurtov, Alia, 4–7

Hawks, John, 12
Homo antecessor, 52
Homo denisova, 78, 80, 81, 83
Homo floresiensis, 13, 61–62, 64
Homo heidelbergensis, 13, 52, 99
Homo naledi, 10–12, 13
Homo neanderthalensis, 13, 41, 42, 43, 44, 48, 52, 53, 80–81, 83
Homo rudolfensis, 13
Homo sapiens, 10–12, 13, 41, 48, 52, 61, 62, 78, 80, 81, 83, 94, 99
Hoyo Negro, 91–95

ice ages, 44, 61, 91, 92, 96
Ichthyosaurus, 19
illegal fossil sales, 34
Indiana Jones, 28–29
Indonesia, 56–65
Italy, 38–45

Jebel Irhoud, 10
jewelry, 6, 44, 80
Jurassic Coast, 14–19, 22–25

Komodo dragon, 58

Lagar Velho archaeological site, 83
Lamalunga Cave, 40–44
Liang Bua, 58–62, 64
limestone, 7, 17, 25, 38, 40–41, 46, 49, 50, 61, 78, 88, 90
Loch Ness monster, 73
Luna, Pilar, 92
Lyme Regis, England, 18, 22–23, 25

mastodons, 92, 94
Mexico, 88–99
Mongolia, 26–37
Mongolian Academy of Sciences, 31

Naia, 94–96, 99
Nava, Alberto, 88–90, 91, 92, 96
Norell, Mark, 31, 33, 34

Ostrom, John, 34
Oviraptor, 30, 33–34
Owen, Richard, 23

Pääbo, Svante, 79, 81, 83–85
Piltdown Man, 45
Plesiosaurus, 22
Portugal, 83
Protoceratops, 29, 30, 31, 34

Rampasasa people, 64
Reich, David, 76–78, 81
removing fossils, 31
Rising Star cave system, 7, 9, 10–12
Russia, 18, 76, 78–79, 81
Russian Academy of Sciences, 79

Saptomo, Wahyu, 56
sedimentary rock, 20
Siberia, 76–85
Sima de los Huesos (Pit of Bones), 49–50, 52–53
South Africa, 4, 7
spoken language, 53
Stringer, Chris, 52

Tarus, Benyamin, 58
3D scanning, 7, 43, 64, 70
Torres, Trinidad, 46–49
trace fossils, 20

underwater archaeology, 90

Vega Island, 69–72, 74
Vegavis iaai, 72
Velociraptor, 34
volcanic eruptions, 44, 73, 74

Yucatán Peninsula, 90–91, 95, 96

ABOUT THE AUTHOR

Tristan Poehlmann writes nonfiction on history, science, and art. A former museum exhibit developer, he holds a master's degree in writing for children and young adults from Vermont College of Fine Arts. He lives in the San Francisco Bay Area.